AF904549

APPLIED PHYSIOLOGY OF EXERCISE
LABORATORY MANUAL

APPLIED PHYSIOLOGY OF EXERCISE
LABORATORY MANUAL

Dr Govindasamy Balasekaran
Nanyang Technological University, Singapore

Dr Visvasuresh Victor Govindaswamy
Concordia University, USA

Lim Ziyuan Jolene
Nanyang Technological University, Singapore

Boey Peck Kay Peggy
Nanyang Technological University, Singapore

Ng Yew Cheo
Singapore University of Social Sciences, Singapore

World Scientific

NEW JERSEY • LONDON • SINGAPORE • BEIJING • SHANGHAI • HONG KONG • TAIPEI • CHENNAI • TOKYO

Published by

World Scientific Publishing Co. Pte. Ltd.
5 Toh Tuck Link, Singapore 596224
USA office: 27 Warren Street, Suite 401-402, Hackensack, NJ 07601
UK office: 57 Shelton Street, Covent Garden, London WC2H 9HE

Library of Congress Cataloging-in-Publication Data
Names: Balasekaran, Govindasamy, author.
Title: Applied physiology of exercise laboratory manual / Dr. Govindasamy Balasekaran, Nanyang Technological University, Singapore [and four others].
Description: Hackensack : World Scientific, 2022. | Includes bibliographical references and index.
Identifiers: LCCN 2021013060 | ISBN 9789811232794 (hardcover) | ISBN 9789811234163 (paperback) | ISBN 9789811233777 (ebook) | ISBN 9789811233784 (ebook other)
Subjects: LCSH: Exercise--Physiological aspects--Handbooks, manuals, etc.
Classification: LCC QP301 .B313 2022 | DDC 612.7/6--dc23
LC record available at https://lccn.loc.gov/2021013060

British Library Cataloguing-in-Publication Data
A catalogue record for this book is available from the British Library.

Copyright © 2022 by World Scientific Publishing Co. Pte. Ltd.

All rights reserved. This book, or parts thereof, may not be reproduced in any form or by any means, electronic or mechanical, including photocopying, recording or any information storage and retrieval system now known or to be invented, without written permission from the publisher.

For photocopying of material in this volume, please pay a copying fee through the Copyright Clearance Center, Inc., 222 Rosewood Drive, Danvers, MA 01923, USA. In this case permission to photocopy is not required from the publisher.

For any available supplementary material, please visit
https://www.worldscientific.com/worldscibooks/10.1142/12175#t=suppl

Desk Editor: Lai Ann

Typeset by Stallion Press
Email: enquiries@stallionpress.com

This book is dedicated to all exercise professionals including professors, lecturers, teachers, personal trainers, coaches, and fitness leaders. Always develop the passion and enthusiasm to spread the magic word of "exercise" to the world by incorporating safe practices and proper physiological principles. The authors would also like to thank the numerous undergraduate and graduate students who graduated from final year projects (FYP), Nanyang Technological University's Undergraduate Research Experience on CAmpus (URECA), Academic Exercise (AE) programmes, and, Master of Science and doctoral studies under the tutelage of Dr G. Balasekaran, who may have contributed in one way or another to make this book complete with the latest research innovations. Special thanks to Ms Yow Chea Nuan, who is always willing to help for a good cause.

Foreword

J. Larry Durstine, Ph.D., FACSM, FAACVPR, FNAK
Distinguished Professor Emeritus
Department of Exercise Science
The University of South Carolina
Columbia, SC 29208
803-413-6009

Past President
American College of Sports Medicine (ACSM)

Ldurstin@mailbox.sc.edu

"Regular physical activity is one of the most important things people can do to improve their health. Moving more and sitting less have tremendous benefits for everyone, regardless of age, sex, race, ethnicity, or current fitness level. Individuals with a chronic disease or a disability benefit from regular physical activity, as do women who are pregnant. The scientific evidence continues to build — physical activity is linked with even more positive health outcomes than we previously thought. And, even better, benefits can start accumulating with small amounts of, and immediately after doing, physical activity."

Taken from the opening statement of the
Physical Activity Guidelines for Americans
2nd edition, 2018

For many years, exercise professionals have known and provided the scientific evidence that regular physical activity and exercise produce numerous functional and health benefits. Society now realizes and accepts that regular physical activity and exercise are a most important component of good health management. Whether undertaking aerobic and/or resistance exercise training and when coupled with other lifestyle strategies such as a healthy diet and stress management, these lifestyle approaches work synergistically in developing a wide range of functional and health benefits. Physicians and allied health professionals better understand these benefits and approve the use of exercise evaluation in the prevention, diagnosis, and the role of exercise programming in the primary, secondary, and tertiary prevention for many clinical conditions and chronic health problems.

Valuable information regarding athlete/client/patient physiologic functions are provided when assessments of muscular strength, endurance, flexibility, and cardiovascular functional capacity are completed. This information enables exercise professionals to determine a client's overall functional capability, assess safety for physical exertion, prescribe exercise interventions, and evaluate the impact of these interventions. Exercise testing also provides morbidity and mortality prognostic information from hemodynamic, electrocardiographic, cardiovascular, and pulmonary exercise responses. The information provided allows exercise professionals to develop athletes and clients' personalized exercise interventions which is a key benefit to exercise testing. Personalized interventions result in fewer complications and more favorable outcomes than a generic intervention. Favorable intervention health outcomes include a positive altered blood lipid and lipoprotein profile (reduced blood triglyceride and increased high density lipoprotein cholesterol), reductions in blood pressure (especially in persons with high blood pressure), enhanced glucose tolerance, insulin sensitivity, and bone mineral density. Added information concerning functional and health benefits are gained when exercise testing assesses ventilatory threshold and aerobic capacity.

Since the inception of the American College of Sports Medicine nearly 70 years ago, a primary goal of this organization has been to gain a better understanding for using exercise testing and exercise training for healthy persons and for persons with chronic health conditions. Because diseases or disabilities have unique characteristics within each condition, a wide range of functional capacity reductions and overall health risks exist that are determined by factors such as progression of the disease, response to treatment, and presence of concomitant illnesses and disabilities. Information gained from exercise testing provide clinicians and practitioners with new perspectives for the use of advanced exercise testing to develop exercise interventions. Considering that a downward spiral toward exercise intolerance is possible for many people with a chronic disease or disability, the loss of functional and exercise capacity results in diminished self-efficacy, greater dependence on others for daily living, depression, and a reduced capability for normal social interaction. Given the well-established benefits of physical activity and exercise for healthy individuals, regular physical activity and exercise training can provide similar health and functional benefits for individuals having many different chronic diseases and/or disabilities.

The goals of a personalized exercise intervention for individuals with a chronic disease or disability are similar to the goals of a healthy individual. These goals are based on the results of physical assessments and exercise testing information that facilitate the reduction in sedentary lifestyle which will exacerbate disease-specific consequences and reduce functional capacity. Individuals with a disease or disability can easily experience positive health outcomes when appropriate medical interventions and lifestyle changes are combined within a medical management plan aiming to improve functional capacity or merely prevent further physical deterioration. Regardless of the individual's exercise goals, one constant objective for all chronically ill persons is to optimize the overall medical management plan and help the individual achieve greater independence and improved quality of life.

Dr Balasekaran, whom I have known for many years, has asked me to introduce this new Applied Physiology of Exercise Laboratory Manual. This guide provides exercise laboratory practitioners and clinicians with the necessary measurement protocols and practical applications for 15 different laboratory sessions. These sessions are applicable to athletes, healthy individuals, and individuals with chronic health conditions. Topics found in this manual include body composition measurement; weight management; resting metabolic rate measurement; effort measurement; exercise validation trial using the OMNI RPE scale; self-regulation of exercise intensity; oxygen kinetics — maximally accumulated oxygen deficit (MAOD); post-exercise oxygen consumption; steady state vs. maximal exercise; lactate threshold and Wingate anaerobic test; field implementation for lactate threshold and aerobic/anaerobic interval training; prediction of running performances — running energy reserve index (RERI); plasma volume changes during exercise; postprandial lipemia; and the measurement of running economy. Each laboratory session pertains to the measurement of a specific physiologic function and follows a similar topical format; introduction presenting background information to the session, various methods/protocols of measurement, reference tables for measurement values, forms to use during assessment, and questions to enhance understanding.

The content within this comprehensive manual is based on the scientific evidence for the development of athletic performance and clinical applications for chronic health conditions. The information and protocols in this manual will assist any qualified clinician or practitioner in providing safe and effective physical assessment and exercise testing for developing individualized exercise programming for athletes, healthy clients, and patients with chronic health conditions. This book can be used in preparing advanced undergraduate and graduate students for certifications offered by the American College of Sports Medicine. This manual can also serve as a reference for health professionals, including physicians, physician assistants, nurses, physical and occupational therapists, rehabilitation specialists, and exercise practitioners working with athletes, healthy individuals, and individuals suffering from chronic health conditions or disabilities.

Preface

G. Balasekaran, Ph.D., FACSM
Associate Professor
Physical Education & Sports Science
National Institute of Education
Nanyang Technological University
Singapore

Health Fitness Director™
American College of Sports Medicine (ACSM)

President
Asian Council of Exercise & Sports Science (ACESS)

(65) 6790 3686
govindasamy.b@nie.edu.sg

Practical applications of physiology of exercise factual materials found in the *Applied Physiology of Exercise* textbook by my co-authors and me are of paramount importance to understand the principles of training. The *Applied Physiology of Exercise Laboratory Manual* complements the *Applied Physiology of Exercise* textbook where practical applications in both laboratory and field settings are shared in the former. These practical applications are mostly through personal research at the Nanyang Technological University, National Institute of Education, and Human Bioenergetics Laboratory in Singapore. The uniqueness of the laboratory sessions found in the manual attest to the many hours of hard laboratory research work. For example, the Running Energy Research Index (RERI) laboratory was born as a result of a 10-year long research. This laboratory research, like the other researched laboratory sessions in the manual, are then used in practical sessions in physiology of exercise classes to fine-tune the best possible learning experiences for the students. After a long process of fine-tuning and constructive feasibility, the laboratory sessions became concrete and designed specifically for this manual.

Practical applications are important as applying the principles of physiology of exercise is where the coach, teacher, or health practitioner benefit the most. To train an athlete for any sport, the trainer needs to have the relevant data either through a laboratory or field session. Only then he/she can be confident to prescribe exercise workouts and training sessions to help the athlete reap the physiological adaptations through conditioning and/or fine-tuning work. The guessing game of "if" an athlete benefitted from an exercise training session to induce a specific physiological adaptation may become obsolete. Unfortunately, such knowledge is imperative in the exercise training realm. Long gone are the guessing games of trial-and-error training methods of legends like Emil Zatopek or Arthur Lydiard where they tested physiological benefits by exhausting their bodies to the maximum training exercise load. It has long been understood that overtraining may not benefit physiological systems or adaptations in a human body and may cause injuries or burnout. The law of diminishing returns too sets in due to excessive overtraining and the body does not respond to physiological adaptations anymore and only depletes the body of the much-needed energy and may lead to the eventual wearing out of ligaments/muscles which may increase the inevitability of a risk of a permanent injury or poor performance.

I had warmly invited Professor Larry, a dear friend whom I have known for a long time, to write his foreword for this laboratory manual. Prof Larry is a world-renowned expert especially in the area of exercise physiology and was the past President of the American College of Sports Medicine. With his valuable insights, I am honored to have him write the foreword for us.

Hope you enjoy the practical physiological laboratory or field sessions in the *Applied Physiology of Exercise Laboratory Manual* and always refer to the *Applied Physiology of Exercise* textbook to better understand the physiological rationale for such sessions as both books complement each other. Immerse yourself in the learning and we hope you improve in the knowledge of understanding the human body when it undergoes exercise stress. Enjoy training by training intelligently.

I have always believed in these 2 personal quotes of mine: "You don't have to always run fast to be fast" and "When to run fast to be fast". Both quotes work in tandem for training. The need to know “when to run fast” is where proper conditioning is required, and this book will expand the physiology knowledge of an individual so that he/she will be able to distinguish specific training periodisation and recognise the various stages of training.

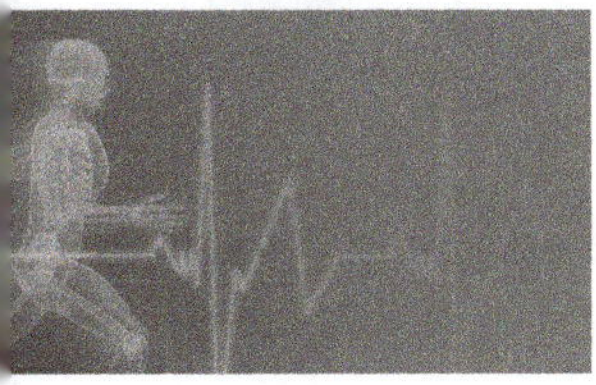

About the Authors

Main Author

Dr Govindasamy Balasekaran, better known as Dr Bala, earned his PhD at the University of Pittsburg under the tutelage of Professor Robert J. Robertson, famous worldwide for his rate of perceived exertion work in exercise, and Dr Silva Arslanian, MD, a well-known pediatric endocrinologist from the University of Pittsburgh Medicine Center, Children's Hospital of Pittsburgh, USA. Following which, he completed his Post-Doctoral Fellowship in molecular genetics at the University of Pittsburgh with Dr Robert E. Ferrell, a world-renowned genetics professor. Dr Bala is the former Head of Physical Education & Sports Science and former Programme Director of Sport Science and Management at the Physical Education and Sports Science department in the National Institute of Education (NIE), Nanyang Technological University (NTU), Singapore. He is currently an Associate Professor at NIE, and his research projects include physiological responses to exercise and adaptations to health and sports performance. In addition, his interests are in investigating the influence of genetic factors on exercise-related outcomes and is currently involved in examining physiological predictors of human performance.

Dr Bala has extensive teaching experience in topics that include anatomy and physiology; human physiology; human functional anatomy; applied exercise physiology; physiological bases of exercise; measurement and evaluation; metabolic and cardio-respiratory aspects of exercise;

nutrition; obesity; health and fitness; hockey; track and field; and fitness and conditioning. He is a certified American College of Sports Medicine Health Fitness Director and a Fellow of the American College of Sports Medicine, and sits on a number of international boards and holds important positions in Asian and global societies such as the Asian Association of Sports Management; Asian Society of Kinesiology; Asian Society of Young Children; Asian Council for Health, Physical Activity & Fitness; The Foundation for Global Community Health; and Federation International D'Education Physique. He is also the current President of the Asian Council of Exercise and Sport Science. He was also the Secretary General of the International Conference for Physical Education & Sports Science (ICPESS) 2010 in conjunction with the 1st Youth Olympics Games organised in Singapore. Furthermore, he was the Executive President of the Asian Society of Young Children as he organised the conference in Singapore. In addition, he is an elected member of the prestigious Sigma Xi, and a member of the European College of Sports Science, The Obesity Society, and American Physiological Society. Moreover, he also conducted numerous workshops for the American College of Sports Medicine certification in health and fitness in Singapore. He also organised the prestigious International Association of Athletics Federations (IAAF) (currently known as World Athletics)-NIE/NTU Chief Coach Youth Academy in Singapore for elite coaches from all around the world. With many first-rate published research papers, proceeding papers, books, and book chapters in the area of sport science, Dr Bala serves as the editorial board member for a number of international journals and has been invited as keynote speaker/invited speaker/presenter for various renowned conferences. He is also currently involved in implementing HOPSports® Inc. Brain Breaks® in Singapore schools and engaging in collaborative global research on Brain Breaks® for children to enjoy physical activity during classroom and physical education lessons. He has won the NTU Nanyang Award (School) and several NIE Commendation of Teaching awards. The NTU Healthy Lifestyle Committee, chaired by him, won the Singapore HEALTH Award (Platinum and Gold) given by the Singapore Health Promotion Board. NTU is the only university that was awarded the Platinum award among all corporate organisations. He also stays on campus as a faculty of residence since 2016 to interact and help local and foreign students to enjoy and live a holistic campus life. He is one of the faculty who helps run "Spartans", which is an exercise programme catering to in-house student residents. It has become one of the favourite programmes for student residents and is also expanded and available online to cater to more students, allowing them to enjoy exercising and to take a breather from their hectic campus life. He also volunteered as a committee member in the organisation of the Ministry of Home Affairs (Police) Real Run for 15 years, the Ministry of Education Olive Run for 12 years and counting, and the Kebun Bahru Link Resident Committee 5-km run for many years till 2015. These running events were organised for civil servants and the public, which started off as a small-scale event but eventually became grand and popular.

From a school boy athlete (hockey, track and field, and cross-country) to a performance athlete (track and field, road races, and cross-country) who had represented his country (Singapore) in long-distance running events, he had won medals in various international and local meets. He had also earned the distinction of having qualified and raced in the prestigious National Collegiate Athletic Association (NCAA) cross-country championships in the USA. Dr Bala holds both Level I and Level II IAAF (currently known as World Athletics) coaching certificates. A team manager at the 2009 and 2015 Southeast Asian (SEA) Games, as well as the 2016 Youth SEA Games, he has coached many national and local schools' long-distance athletes who went on to achieve national honours and established national and local school records. He also volunteered and was the Assistant Honorary Secretary (2014–2016 June) and Vice President of training and selection (2016–2018 November) for the national Singapore Athletics Association. Dr Bala currently provides voluntary services to Singapore athletes as their national coach. He is currently the President of Cougars Athletic Association, an associate affiliate with Singapore Athletics, which is a non-profit club that nurtures young athletes for the future. Dr Bala has vast experience in the area of athletics coaching and is interested in vast aspects surrounding performance in track and field including sports science in various sports.

Co-Authors

Dr Visvasuresh Victor Govindaswamy

Dr Victor graduated with a Master of Science in Computer Science and Engineering and a Bachelor of Science in Electrical and Computer Engineering from the University of Texas in Austin, USA. He has a PhD from the University of Texas in Arlington and is currently an Associate Professor, Director of Computer Science Programs at Concordia University, Chicago, USA. He has worked on a number of research projects, a notable one being in the area of preparing dependable, dynamic real-time application systems for an adaptive resource management environment, which was sponsored by the Defense Advanced Research Projects Agency (DARPA), USA, the central research and development organisation for the Department of Defense and the National Aeronautics and Space Administration (NASA). Dr Victor has a strong interest in physiological research and has collaborated on a number of research projects. He is also an avid runner and has participated in many races during his school and university days.

Jolene Lim

Jolene graduated from Loughborough University, United Kingdom, with a Master of Science in Sport and Exercise Science and worked as a research assistant at the Nanyang Technological University, Singapore. She is currently pursuing her PhD in Sports Science at the Nanyang Technological University and presently working at Singapore Shooting Association. She volunteers as Vice President Development at Cougars Athletic Association. She also deals with research associated with applied sports science in children, youth, and adults, and is keenly involved in exploring the physiological aspects of human performance. She is also a sports enthusiast and engages in regular physical activities.

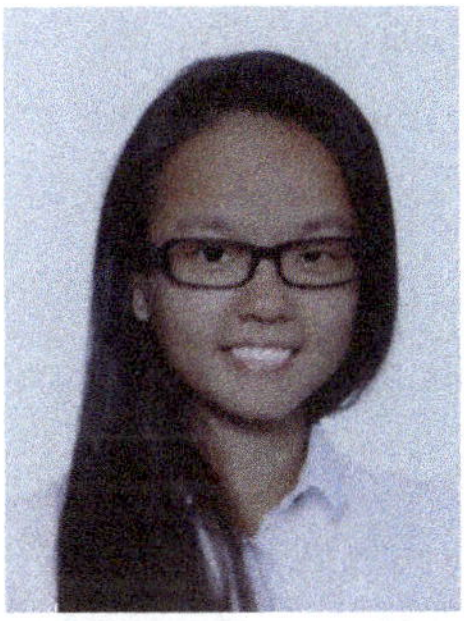

Peggy Boey

Peggy is a Physical Education teacher and is currently pursuing her Master of Science from the National Institute of Education, Nanyang Technological University. She graduated with a Bachelor in Physical Education & Sports Science at the National Institute of Education. She is a member of the American College of Sports Medicine and a life member of the Asian Council of Exercise & Sports Science. She has been involved in intensive research on children exercising within a safe intensity using the OMNI Rate of Perceived Exertion Scale and has implemented it in her teaching of Physical Education at her school. Currently, she is involved in human performance research. She also volunteers as Vice President Competitions Organising at Cougars Athletic Association. Peggy is a former competitive swimmer and has competed in long-distance races in Singapore. She also holds the Level I International Association of Athletics Federations (IAAF) coaching certificate and is looking forward to coaching younger athletes in track and field.

Ng Yew Cheo

Yew Cheo is a national athlete who obtained her Diploma in Physical Education & Sports Science at the National Institute of Education, and is a degree holder in Exercise Science from the Singapore University of Social Sciences. She is also a member of the American College of Sports Medicine and a life member of the Asian Council of Exercise & Sports Science. Furthermore, she is a Future Leader Volunteer of The Foundation for the Global Community Health, promoting physical activity in the community. She is involved in assisting the implementation of HOPSports® Inc. Brain Breaks® in Singapore schools and engaging in collaborative global research on Brain Breaks® for children to enjoy physical activity during classroom and physical education lessons. She is a former Physical Education teacher and is actively involved in track and field and trains younger athletes to reach their potential. She has represented the nation and won medals at international and national meets and is currently running competitively. She also coaches voluntarily and volunteers as Honorary Secretary at Cougars Athletic Association. On top of that, she works on research projects engaging children and adolescents in using the OMNI Rate of Perceived Exertion Scale during exercise and Physical Education lessons. Currently, she is involved in human performance research. Additionally, she holds the Level I International Association of Athletics Federations (IAAF) coaching certificate.

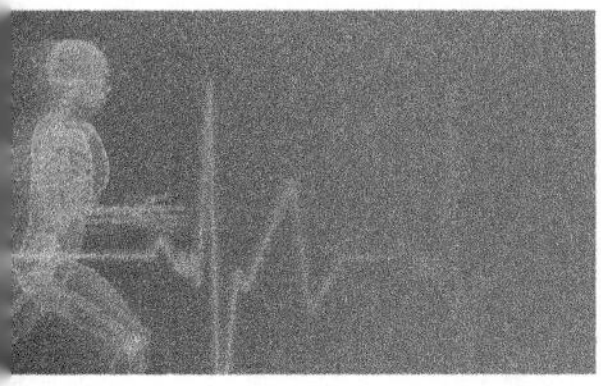

Acknowledgements

The main author, Dr G. Balasekaran, would like to give special thanks to Professor Robert J. Robertson (PhD, Physiology of Exercise) and Professor Robert E. Ferrell (PhD, Molecular Genetics), from University of Pittsburgh for their utmost, invaluable advice and guidance during his PhD and post-doctoral academic journey, respectively, which eventually led to the development of this applied physiology laboratory manual textbook. Additionally, he would also like to specially thank Dr Silva Arslanian, MD (Endocrinologist), from the Children's Hospital of Pittsburgh, whom he worked with on research projects during his doctoral days. Many thanks to his professors, Professor James G. Mill, Professor Elaine Blair, Professor Archie Moore, and Professor Edward Sloniger of Indiana University of Pennsylvania, Department of Kinesiology, Health, and Sport Science for their mentorship, guidance, and support, and teaching their invaluable knowledge during his masters and undergraduate days. A special thanks to Professor J. Larry Durstine, past President of the American College of Sports Medicine (ACSM) and Distinguished Professor Emeritus in the Department of Exercise Science, University of South Carolina, for writing the foreword for the book.

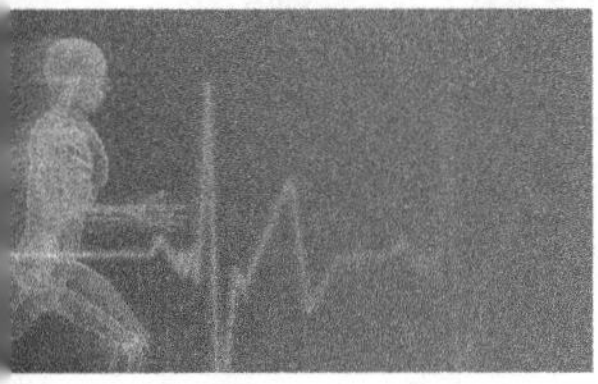

Contents

A SHORT HINT

Here are some hints on how to use this laboratory manual. Users can choose what kind of laboratories they would like to use during a semester, or they can spread the laboratories over 2 semesters, combining theory and practice. Choices can also be made from easier laboratories for introductory content or a harder laboratory for senior classes. The book will complement most physiology of exercise syllabuses from introductory level to higher ability courses as it is comprehensive and covers a wide range of topics. Also, there are laboratory questions for every chapter with marks allocated. The marks allocated can be changed according to the user's preference and can be used in assessment components for passing the module. The students must interpret the results/data of the laboratory sessions to the best of their knowledge and use references from the literature that can be included to substantiate their results or findings. All references used have to be included in the laboratory report if the user is marking and giving feedback to the students. In addition to the diagrams required by the questions, additional diagrams may be used to illustrate the results. Look at the marks allocation closely for each question as this is a guide on how much you should write for each question.

Enjoy the joys of practical physiology!

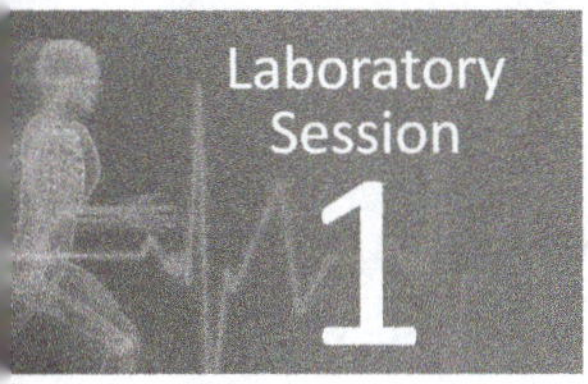

Body Composition: Quantifying Fat Mass

Body composition is defined by the American College of Sports Medicine (ACSM) (2018) as the "relative proportion of fat and fat-free tissue in the body." It is also listed as a component of the health-related fitness and where should all targeted components would improve overall health status and prevent illness related to sedentary lifestyle (Figure 1).

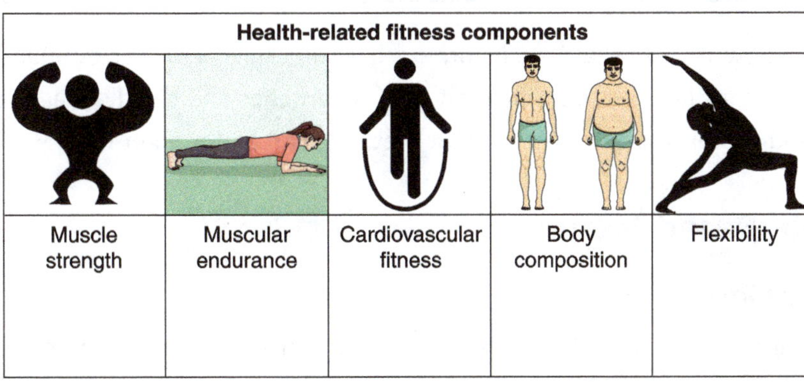

Figure 1. Diagram of health-related fitness components.

Body composition measurement has emerged from a single compartment of weight measurement to a multiple-compartment model that consists of water, mineral, fat, and protein (Balasekaran, 2003; Balasekaran, Gupta, & Govindaswamy, 2010). The basis of a two-compartmental body composition measurement includes fat mass and fat-free mass (Gupta et al., 2011). High fat mass is often one of the factors that leads to cardiovascular diseases such as coronary artery diseases, non-insulin dependent diabetes mellitus, hypertension, certain types of cancers, and hyperlipidemia. Understanding the proportion of both fat and lean muscle mass can help individuals better understand the changes that their body undergo, especially during weight management (Danadian et al., 1999; Balasekaran, 2003; Mayo, Grantham, & Balasekaran, 2003; Balasekaran et al., 2005, 2015, 2018). In sports, excess fat hinders performance as it does not contribute to muscular force production (Balasekaran et al., 2005; Gupta et al., 2011; Balasekaran, Mayo, & Lim, 2019). Moreover, it is additional weight that the individual needs to carry to perform the exercise, and thus requires more energy to burn, which may hinder optimal performance (Balasekaran & Loh, 2009; Balasekaran et al., 2014).

Expression of fat weight
Percent body fat = (fat weight/total body weight) $\times$ 100

Below are some field and laboratory base measurements and evaluation methods for fat mass classification. The advantages and disadvantages of all body composition measurement methods can be found in Table 9.

Field-Based Methods

Body Mass Index

A measurement of body weight in kilograms divided by height in metres squared ($kg{\cdot}m^{-2}$) (Table 1). Individuals with more fat in the trunk section, especially abdominal fat, are at higher risk of heart diseases compared to those who have equal amount of fat but have more of it in their extremities.

Table 1. World Health Organization (WHO) body mass index (BMI) classifications of general population and Asians (adapted from BMI, 1998; WHO, 2004).

	WHO General Population BMI Classifications ($kg{\cdot}m^{-2}$)	WHO Asian BMI Classifications ($kg{\cdot}m^{-2}$)
Underweight	> 18.5	> 18.5
Normal	18.5 – 24.9	18.5 – 22.9
Overweight	25.0 – 29.9	23.0 – 24.9
Obese	$\geq$ 30.0	$\geq$ 25

Values are in body mass index ($kg{\cdot}m^{-2}$).

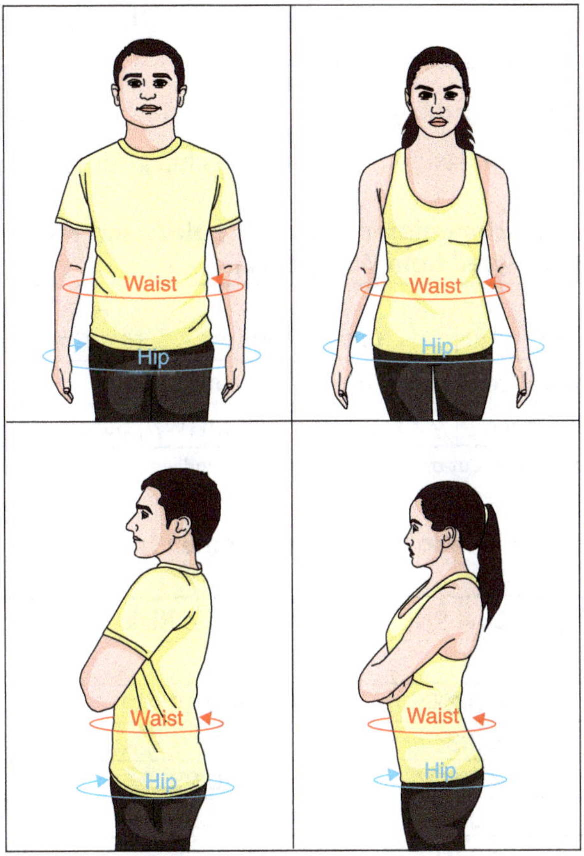

Figure 2. Measurement of circumference of waist and hip for the waist-hip ratio.

Waist-Hip Ratio

A simple measurement via circumference of the waist divided by the circumference of the hips (Figure 2, Tables 2.1 & 2.2). Those who are at an increased risk of obesity are those with ratio above 0.90 for men and 0.85 for women (Table 2.2 refers to age stratified for waist-hip ratio norms).

Procedure (WHO, 2011) (Figure 2):

1. Participant will stand with hands at the side with weight evenly distributed on both feet and legs positioned close together
2. Measurements will be done on two sites for the waist and hip girth

  - Waist girth is the narrowest part of the torso below the rib cage and above the umbilicus
  - Hip girth is the maximum circumference of the buttocks
3. Waist girth measurement divided by hip girth measurement

Table 2.1 Waist circumference cut-off points and risk of metabolic complications (adapted from WHO, 2011).

Indicator	Cut-Off Points	Risk of Metabolic Complications
Waist circumference	> 94 cm (M); > 80 cm (W)	Increased
Waist circumference	> 102 cm (M); > 88 cm (W)	Substantially increased

M: men; W: women. Note: Cut-off points differ for older individuals.

Table 2.2 Waist-hip ratio for men and women (age stratified) (adapted from ACSM, 2014).

Age (Year)	Low Risk	Moderate Risk	High Risk
		Men	
20–29	< 0.83	0.83–0.88	> 0.88
30–39	< 0.84	0.84–0.91	> 0.91
40–49	< 0.88	0.88–0.95	> 0.95
50–59	< 0.90	0.90–0.96	> 0.96
60–69	< 0.91	0.91–0.98	> 0.98
		Women	
20–29	< 0.71	0.71–0.77	> 0.77
30–39	< 0.72	0.72–0.78	> 0.78
40–49	< 0.73	0.73–0.79	> 0.79
50–59	< 0.74	0.74–0.81	> 0.81
60–69	< 0.76	0.76–0.83	> 0.83

Norms for waist-hip ratios are categorised according to age, gender, low risk, moderate risk, and high risk.

Skinfold Test

Based on the principle that the amount of subcutaneous fat is directly proportional to the total amount of body fat, the skinfold test is used to measure the amount of fat mass from the sum of specific sites. The body fat percentage prediction is relatively accurate if performed properly by a trained technician using a high-quality skinfold caliper (Figure 3). The proportion of subcutaneous to total fat varies with gender, age, race, ethnicity, and other factors (ACSM, 2012).

The accuracy of body composition from regression equation depends on securing accurate measures of skinfold fat.

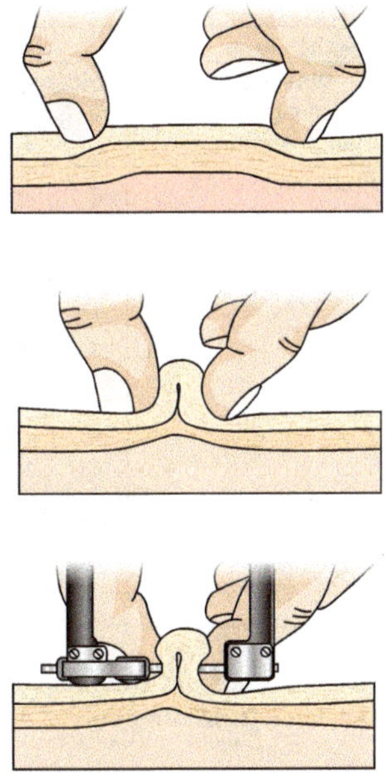

Figure 3. Measuring body fat using skinfold.

Procedure for measurement (Figure 3):

1. Take all measurements from the right side of the body
2. Grasp the skinfold by the thumb and index finger. The caliper is perpendicular to the fold — approximately 1 cm from the thumb and forefinger
3. Release the caliper grip slowly so that full tension is exerted on the skinfold
4. Read the dial to the nearest 0.5 mm for approximately 1–2 seconds after the grip has been released
5. Take 3 measurements at each site and average these 3 numbers

Below is a list of pointers to take note of during measurement:

- Measurements should be taken from the first site to the last site for 3 repetitions (non-consecutively)
- Take measurements when the skin is dry
- Do not take measurements immediately after a participant has exercised (after exercise, there is fluid loss below the skin that may decrease the thickness of skinfold due to some dehydration. The moist skin also makes it difficult to grasp the skinfold.)
- Practice ensures accurate measurement of skinfold fat

Skinfold measurements can be done either on 3 or 7 sites (Jackson, Pollock, & Gettman, 1978) (Figure 4).

	Men	Women
3 – sites area	chest, abdomen, and thigh	triceps, suprailiac, and thigh
7 – sites area (same sites for men & women)	triceps, chest, midaxillary, subscapular, suprailiac, abdominal, and thigh (6) (2) (1) (4) (5) (3) (7)	
Anatomical sites	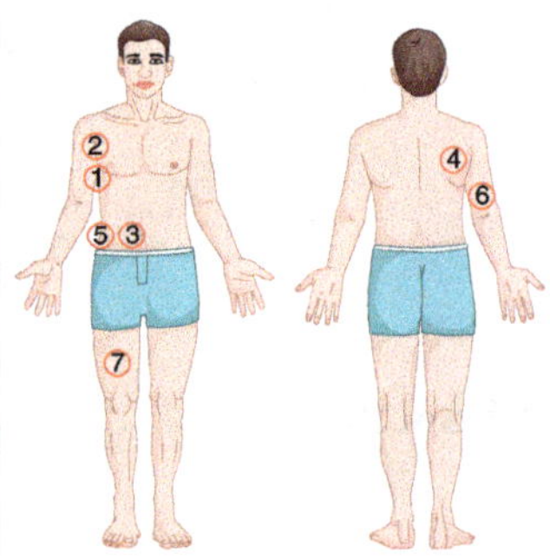	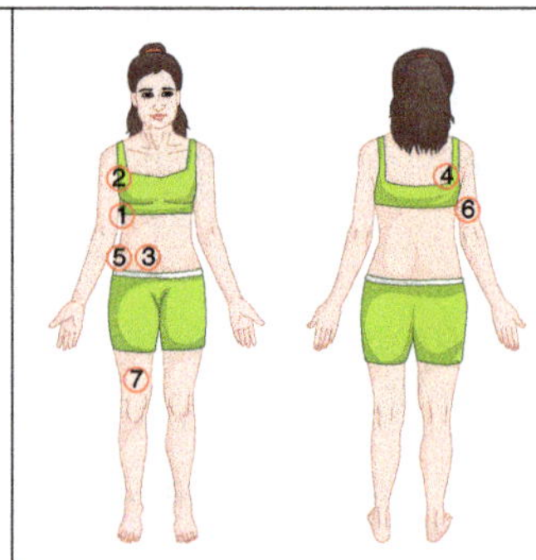

Figure 4. Skinfold site measurements for men and women. Note: You may refer to ACSM Guidelines (2018) for detailed descriptions of skinfold site measurements for further accuracy.

Analysing results for skinfold measurement can be done using Pollock's nomogram (Figure 5) or equations (Figure 6 or Table 3). To obtain the body fat percentage from the nomogram, just use a straight edge to connect the participant's age with the skinfold value. The body fat percentage is rated where the straight edge crosses the line representing the participant's gender. Teachers and other practitioners that work with children in schools may also use Figure 6 and Table 3 to calculate children's body fat percent using the skinfold test.

The skinfold prediction equations (Figure 6) were derived from the following equations:

Equation 1:

Siri (1961) converted body density to body fat based on fat-free body density of 1.1 g/cc and fat density of 0.9 g/cc (f = body fat percentage):

$$f = (495/D) - 450$$

Equation 2:

Brozek (1959) recommended that fat can be estimated from body density using the equation based on the chemical composition of a man:

$$f = [(4.57/D) - 4.142] \times 100$$

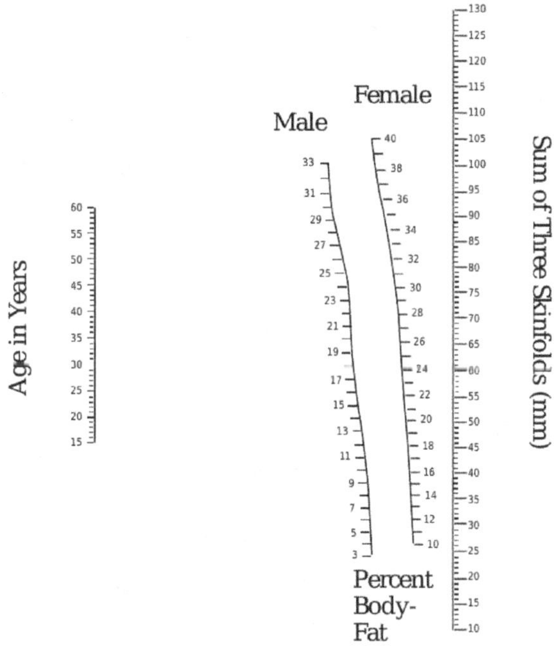

Figure 5. Nomogram to derive percentage of body fat from the total of three skinfolds (adapted from Baun, Baun, & Raven, 1981).

SKF sites	Population subgroups	Equation	Reference
∑7SKF (chest + abdomen + thigh + triceps + subscapular + suprailiac + midaxilla)	Black or Hispanic women, 18–55 years	Db (g/cc)[a] = 1.0970 − 0.00046971 (∑7SKF) + 0.00000056 $(\sum 7SKF)^2$ − 0.00012828 (age)	Jackson et al. (1980)
	Black men or male athletes, 18–61 years	Db (g/cc)[a] = 1.1120 − 0.00043499 (∑7SKF) + 0.00000055 $(\sum 7SKF)^2$ − 0.00028826 (age)	Jackson and Pollock (1978)
∑4SKF (triceps + anterior + suprailiac + abdomen + thigh)	Female athletes, 18–29 years	Db (g/cc)[a] = 1.096095 − 0.0006952 (∑4SKF) + 0.00000011 $(\sum 4SKF)^2$ − 0.0000714 (age)	Jackson et al. (1980)
∑3SKF (triceps + suprailiac + thigh)	White or anorexic women, 18–55 years	Db (g/cc)[a] = 1.0994921 − 0.0009929 (∑3SKF) + 0.0000023 $(\sum 3SKF)^2$ − 0.0001392 (age)	Jackson et al. (1980)
(chest + abdomen + thigh)	White men, 18–61 years	Db (g/cc)[a] = 1.109380 − 0.0008267 (∑3SKF) + 0.0000016 $(\sum 3SKF)^2$ − 0.0002574 (age)	Jackson and Pollock (1978)
∑2SKF (triceps + calf)	Black or white boys, 6–17 years	%BF = 0.735 (∑2SKF) + 1.0	Slaughter et al. (1988)
	Black or white girls, 6–17 years	%BF = 0.610 (∑2SKF) + 5.1	Slaughter et al. (1988)

∑SKF = sum of skinfolds (mm).

[a]Use population-specific conversation formulas to calculate %BF (percent body fat) from Db (body density).

Figure 6. Skinfold prediction equations (adapted from Jackson & Pollock, 1978; Jackson, Pollock, & Ward, 1980; Slaughter et al., 1988).

Table 3. Calculation of body fat percentage from skinfold measurement (adapted from Jackson & Pollock, 1978; Golding, Myers, & Sinning, 1989; Slaughter et al., 1988).

Group	Equation
Adult males	% BF = $0.39287(X_1) - 0.00105(X_1)^2 + 0.15772(X_2) - 5.18845$, where X_1 = sum of abdominal, suprailiac, and triceps skinfolds and X_2 = age
Adult females	% BF = $0.41563(X_1) - 0.00112(X_1)^2 + 0.03661(X_2) - 4.03653$, where X_1 = sum of abdominal, suprailiac, and triceps skinfolds and X_2 = age
Male children and adolescents (8–18 years)	% BF = $0.735(X_1) + 1.0$, where X_1 = sum of triceps and calf skinfolds
Female children and adolescents (8–18 years)	% BF = $0.610(X_1) + 5.1$, where X_1 = sum of triceps and calf skinfolds

Alternatively, an individual can determine their health status from the general body fat percentage categories as shown below (Tables 4 & 5).

Table 4. Body fat percentage for males according to age (adapted from American College of Sports Medicine, 2018).

		Age (year)					
%		20–29	30–39	40–49	50–59	60–69	70–79
99	Very lean	4.2	7.3	9.5	11.0	11.9	13.6
95		6.4	10.3	12.9	14.8	16.2	15.5
90	Excellent	7.9	12.4	15.0	17.0	18.1	17.5
85		9.1	13.7	16.4	18.3	19.2	19.0
80		10.5	14.9	17.5	19.4	20.2	20.1
75	Good	11.5	15.9	18.5	20.2	21.0	21.0
70		12.6	16.8	19.3	21.0	21.7	21.6
65		13.8	17.7	20.1	21.7	22.4	22.3
60		14.8	18.4	20.8	22.3	23.0	22.9
55	Fair	15.8	19.2	21.4	23.0	23.6	23.7
50		16.6	20.0	22.1	23.6	24.2	24.1
45		17.5	20.7	22.8	24.2	24.9	24.7
40		18.6	21.6	23.5	24.9	25.6	25.3

(Continued)

Table 4. (*Continued*)

		Age (year)					
%		20–29	30–39	40–49	50–59	60–69	70–79
35	Poor	19.7	22.4	24.2	25.6	26.4	25.8
30		20.7	23.2	24.9	26.3	27.0	26.5
25		22.0	24.1	25.7	27.1	27.9	27.1
20		23.3	25.1	26.6	28.1	28.8	28.4
15	Very poor	24.9	26.4	27.8	29.2	29.8	29.4
10		26.6	27.7	29.2	30.6	31.2	30.7
5		29.2	30.2	31.3	32.7	33.3	32.9
1		33.4	34.4	35.2	36.4	36.8	37.2

Table 5. Body fat percentage for females according to age (adapted from American College of Sports Medicine, 2018).

		Age (year)					
%		20–29	30–39	40–49	50–59	60–69	70–79
99	Very lean*	11.4	11.2	12.1	13.9	13.9	11.7
95		14.0	13.9	15.2	16.9	17.7	16.4
90	Excellent	15.1	15.5	16.8	19.1	20.2	18.3
85		16.1	16.5	18.3	20.8	22.0	21.2
80	Good	16.8	17.5	19.5	22.3	23.3	22.5
75		17.6	18.3	20.6	23.6	24.6	23.7
70		18.4	19.2	21.7	24.8	25.7	24.8
65		19.0	20.1	22.7	25.8	26.7	25.7
60		19.8	21.0	23.7	26.7	27.5	26.6
55	Fair	20.6	22.0	24.6	27.6	28.3	27.6
50		21.5	22.8	25.5	28.4	29.2	28.2
45		22.2	23.7	26.4	29.3	30.1	28.9
40		23.4	24.8	27.5	30.1	30.8	30.5
35	Poor	24.2	25.8	28.4	30.8	31.5	31.0
30		25.5	26.9	29.5	31.8	32.6	31.9
25		26.7	28.1	30.7	32.9	33.3	32.9
20		28.2	29.6	31.9	33.9	34.4	34.0

(*Continued*)

Table 5. (*Continued*)

	Age (year)						
%		20–29	30–39	40–49	50–59	60–69	70–79
15	Very poor	30.5	31.5	33.4	35.0	35.6	35.3
10		33.5	33.6	35.1	36.1	36.6	36.4
5		36.6	36.2	37.1	37.6	38.2	38.1
1		38.6	39.0	39.1	39.8	40.3	40.2

*It is not recommended for women to fall below 10–13% body fat.

Laboratory Methods

Hydrostatic Weighing

On the basis of Archimedes' principle (completely submerged body — the volume of fluid equates to the volume of the body), this is one of the earlier most popular methods of measuring body density. In the early years in body composition research, this method provided the criterion measurement for validating field methods (e.g. skinfold measurement, etc.) for determining body composition (Figure 7, Tables 6 & 7).

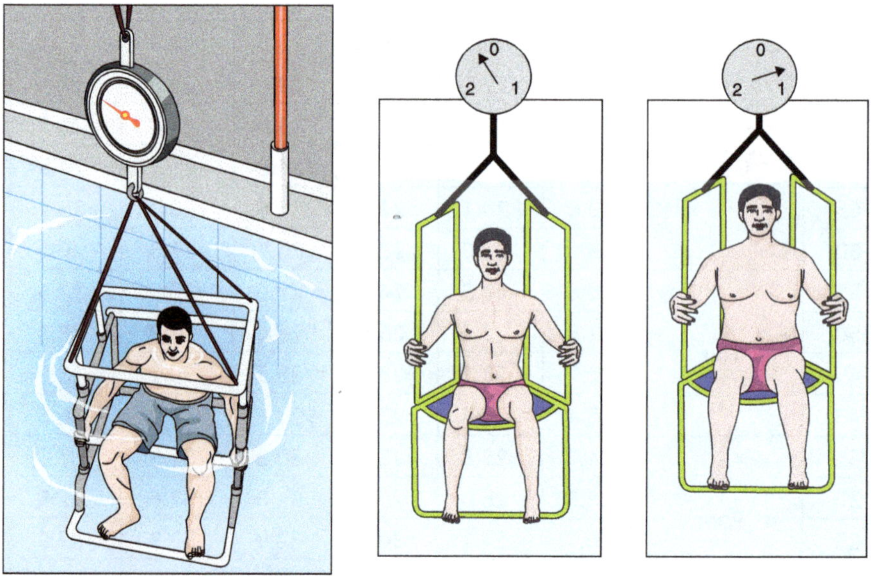

Figure 7. Hydrostatic weighing.

Table 6. Body density sample calculation (adapted from Howley & Franks, 1986).

Body Density
Participant A Gender: Male Age (A): 24 years Body weight in air (WA): 110 kg Body weight in water (WW): 3.5 kg Height: 1.75 m Residual Lung Volume (RLV): −2.968 + 2.78(H/m) + 0.01A + 0.008(W/kg) (Yap, Chan, & Chan, 2001) (Table 7) Water temperature correction factor/water: 0.996 (Howley & Franks, 1986) Density (DB) = WA/(WA − WW/0.996) − RLV 110/((110 − 3.5/0.996) − 3.017) = 110/((110 − 3.514) − 3.017) = 110/((106.486) − 3.017) = 110/103.469 = 1.063 kg/litre or 1.063 g/cc Body density can be converted to body fat percentage using the following equations: Siri equation (Equation 1): Body fat percentage: f = 495/D − 450 f = 495/1.063 − 450 f = 15.66% With reference to Table 4, this participant is classified under Fair at 55%. OR Brozek equation (Equation 2): Body fat percentage: f = [(4.57/D) − 4.142] × 100 f = [(4.57/1.063) − 4.142] × 100 f = (4.299 − 4.142) × 100 f = 0.157 × 100 f = 15.7% With reference to Table 4, this participant is classified under Fair at 55%.

Table 7. Comparison between different prediction equations done by direct measurement (mRV) on residual lung volume (RV) (Paksaichola et al., 2014).

Validation group	RV (l)	mRV – pRV (l)	p-value*	95% CI of diff*
mRV[†]	1.189 ± 0.223			
$RV = -3.236 + 0.024(H/\text{cm}) + 0.028A$[†]	1.189 ± 0.135	0.000 ± 0.177	1.00	−0.032, 0.032
$RV = -3.9 + 0.0813(H/\text{in}) + 0.009A$[‡]	1.296 ± 0.152	−0.107 ± 0.191	0.00	−0.141, −0.072
$RV = -2.307 + 0.057(H/\text{in}) + 0.01A + 0.005(W/\text{lb})$[§]	0.843 ± 0.112	0.346 ± 0.215	0.00	0.307, 0.384
$RV = -2.968 + 2.78(H/\text{m}) + 0.01A + 0.008(W/\text{kg})$[Ψ]	1.175 ± 0.121	0.014 ± 0.199	0.46	−0.022, 0.049
$RV = -4.40778 + 0.03403(H/\text{cm})$[¶]	0.894 ± 0.163	0.295 ± 0.204	0.00	0.259, 0.332

*Paired t-test compared to mRV. Independent variables: H = height, A = age (year), and W = weight.
[†]Obtained from present study. Ethnicity: [†]Thai (Paksaichola et al., 2014), [‡]Caucasian (Goldman & Becklake, 1959), [§]Chinese (Ching & Horsfall, 1977), [Ψ]Singaporean (Yap, Chan, & Chan, 2001), [¶]Japanese (Demura, Yamaji, & Kitabayashi, 2006).
Note: Select equation based on the ethnicity of your country.

Procedure for measurement (Figure 7):

1. Measure dry weight of participant in minimal clothing
2. Participant will sit on a special seat, expel all air from lungs, and continue expelling air until he/she is fully submerged underwater in the water tank
3. Underwater weight is then determined

Bioelectric Impedance

Table 8. Body fat calculations using bioelectric impedance analysis (note: different BIA machines use different calculations and use a variety of prediction equations).

Body Fat (refer to Table 6's calculations for Siri's and Brozek's equations)
Siri equation (Equation 1) Brozek equation (Equation 2)

The bioelectric impedance analysis (BIA) determines various body components such as bone density, body weight, body fat percentage, and lean mass through different electrical volts transmitted into the body (Figure 8). The volts are sent through either the hands and/or feet that are in contact with the electrodes. The machine operates on the principle that the resistance to an applied electric current is inversely related to the amount of lean mass within the body, and the calculations are configured (Table 8).

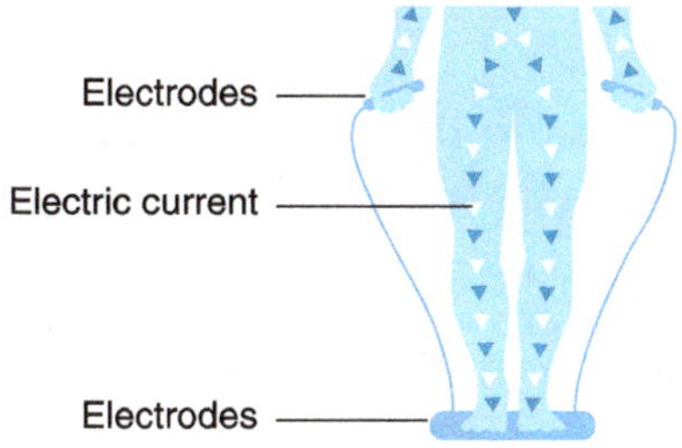

Figure 8. Electric current transmitting through body during bioelectric impedance analysis.

Procedure (Figure 8):

1. Participant will stand on the machine with hands and/or feet in contact with electrodes
2. Hold still and wait for machine to process calculation

Areas for consideration when using BIA:

- Hydration status
- Body temperature
- Time of the day

Air Displacement Plethysmography/BOD POD

Air displacement plethysmography (ADP) is a two-compartment measurement model that calculates body density by assessing mass and body volume (Figure 9). The participant's volume is measured twice or thrice in the BOD POD chamber depending on the consistency of measurement. The body volume is then corrected for thoracic volume before actual results are calculated. The participant will have to wear minimal clothing (i.e. swim wear and swimming cap) before entering the BOD POD chamber. Additionally, the participant must not eat or participate in any vigorous exercise 2 hours before the assessment.

Procedure (Figure 9):

1. Laboratory personnel will calibrate the BOD POD
2. Participant will be dressed in minimal clothing (see above paragraph)
3. Take participant's weight using a very precise, integrated electronic scale
4. Participant will sit in the chamber, relaxed, while laboratory personnel will key in his/her basic anthropometric values and begin the assessment
5. Assessment will take approximately 5 minutes to complete

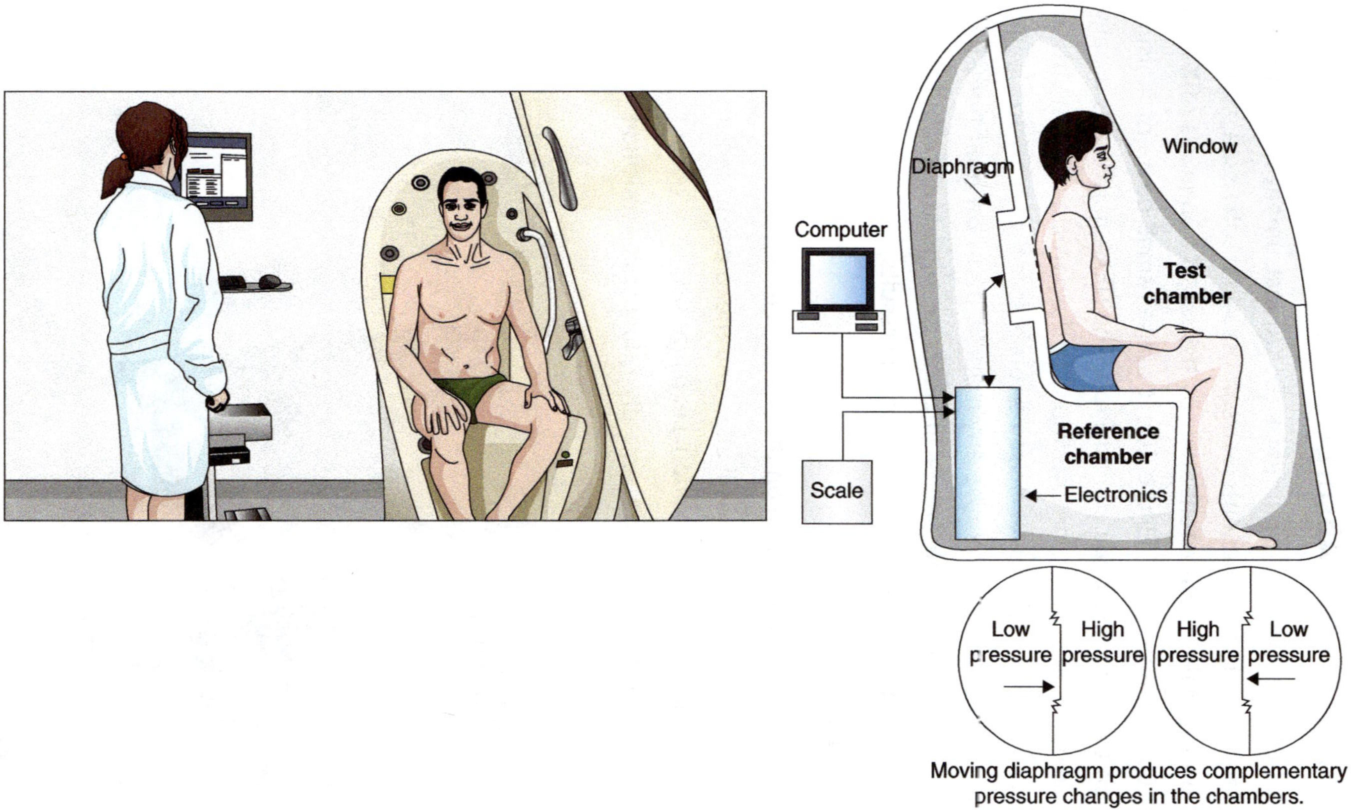

Figure 9. Air displacement plethysmography (ADP) or BOD POD.

Dual-Energy X-Ray Absorptiometry

The dual-energy X-ray absorptiometry (DEXA) instrument differentiates body weight into the components lean soft tissue, fat soft tissue, and bone based on the differential attenuation by tissues of two levels of X-rays (Lee & Balasekaran, 2010; Figure 10). It utilises small doses of ionising radiation to determine body composition. The DEXA machine is usually operated by certified personnel. Additionally, the participant must not eat or participate in any vigorous exercise 2 hours before the scan. Participants are also usually asked to remove any metallic items on their body (i.e. jewellery, rings, earrings, etc., metal implants can remain intact but notify the laboratory personnel as it may affect the results).

Procedure (Figure 10):

1. Participant to lie down on the machine facing upwards
2. Laboratory personnel will align and move participant to ensure body in anatomical position, i.e. palms up
3. Scanning will take approximately 7 minutes to conclude

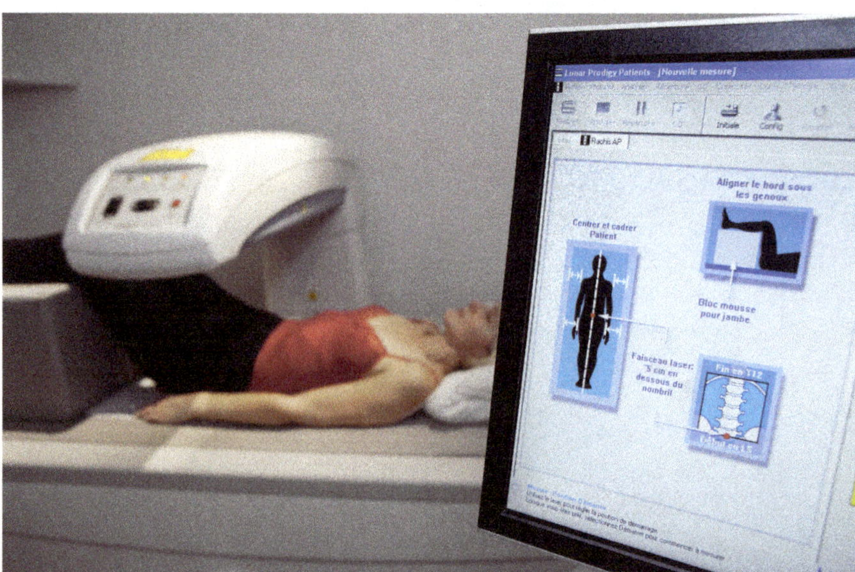

Figure 10. Dual-energy X-ray absorptiometry (DEXA) scan results.

Table 9. Advantages and disadvantages of measurement methods.

	Advantages	Disadvantages	Remarks
Waist to Hip Ratio (WHR)	A simple measure that can be taken at home by people to monitor their own levels. Can be used in mass settings like in schools.	WHR can be measured inaccurately if proper procedures of identifying location of measurement is not administered strictly.	The higher the amount of fat stored around the waist, the greater the risk to health than fat stored elsewhere in the body, which is highly correlated to coronary heart disease risk factors.
Body Mass Index (BMI)	Simple calculation from standard measurements. Can be used in mass settings like in schools.	BMI can be inaccurate, for example, large, muscular, and lean athletes may be classified as high BMI level which incorrectly categorises them as obese.	Other measures of body composition would be preferable if available.
Skinfold measurement	Skinfold measurements are widely utilised to assess body composition. It is a lot simpler than hydrostatic weighing. The tests costs are minimal.	Skinfold measurement can be inaccurate, for example, if the tester "pinches" too little or too much, the results will not be accurate. The tester must also be well-trained in identifying anatomical sites for accurate results. Not suitable for mass testing unless tester tests for 2 to 3 sites. This may reduce accuracy of body composition as subcutaneous fat is distributed	Skinfold test administered after exercise affects measurement. The skill of the tester is also important as reliability is affected.

	Advantages	Disadvantages	Remarks
		all over the body. In mass settings, if there are more sites to measure, the error(s) might be multiplied due to the tester's competency and time constraints. Thus fewer sites (e.g. 3 sites, Figure 4) may be more applicable as the participant will have an approximate body fat percentage value using the nomogram (Figure 5). Alternatively, the tester can replace the skinfold test with another body composition measurement method.	
Hydrostatic Weighing	Underwater weighing in the early years of body composition is the most widely used test of body density. It was the criterion measure for other indirect measures until DEXA became the gold standard for body composition assessments.	The equipment required to do underwater weighing is expensive. The tanks or swimming pools are mostly located at universities or other research institutions, and it is overall not easily accessible for the general population. Also, not many participants can exhale all the air in the lungs and continue exhaling while going underwater. This phobia may be a deterrent to the participants to exhale all air in the lungs, and thus may give inaccurate readings.	Residual lung volume is required for calculations. For more accuracy, it should be measured, though there are calculations for residual volume estimation. An estimation of residual volume is one-third of forced vital capacity.

	Advantages	Disadvantages	Remarks
Bioelectric Impedance Analysis (BIA)	Very simple and quick to analyse body composition. Can be used for mass settings like in schools. Some BIA machines are not expensive, and it is easy to use.	The equipment is relatively expensive (units on the low-end of the scale are available for about USD $70). However, components of measurements may be fewer, with prices ranging up to thousands of dollars depending on what measurements are available. The impedance measurement is affected by body hydration status, body temperature, and time of day, and therefore requires well-controlled conditions to get accurate and reliable measurements.	BIA is based on the principle that the resistance to an applied electric current is inversely related to the amount of fat-free mass within the body.
Air Displacement Plethysmography (ADP) or BOD POD	BOD POD measurements are as accurate as hydrostatic weighing, but quicker and easier to utilise. It can be employed to validate comparison studies with other body composition methods.	The equipment is expensive, and often requires trained laboratory personnel to operate. May not be suitable for mass settings as it is always situated in a laboratory. It may be slightly time-consuming due to the pre-preparations that the participants need to adhere to (i.e. minimal clothing etc.) It may not be suitable for young children to perform.	There are different equations that derive body composition for different ethnicities and children.

	Advantages	**Disadvantages**	**Remarks**
Dual Energy X-ray Absorptiometry (DEXA)	DEXA measurements are accurate and based on a three-compartmental model as compared to two compartments in other methods. It can also distinguish regional as well as whole body parameter of body composition. Some DEXA machines can calculate bone density, which is a good measurement for accessing health status of the bone.	The equipment is expensive, and often requires an expert certified radiology personnel to operate. Not suitable for mass settings as it is always situated in a laboratory and need an expert technician's help. It is also time-consuming to measure.	DEXA is considered the gold standard criterion method of body composition analysis.

Record Sheet: Body Composition

Determine your body composition with the following common methods. You can include other methods like the BOD POD, hydrostatic weighing, etc. If your laboratory possesses such equipment, you can further make comparisons between values obtained.

My Record

Name: ________________________ Age: _____ Gender: Male/Female

1. **Dual-Energy X-Ray Absorptiometry (DEXA)**

   Date of Test: ________________________ Total Body Fat: _________

   Classification: ____

   Bone Mineral Density: ____________ Classification: ______________

2. **Skinfold Measurement** Date of Test: _______________

   Site 1: _____________mm

   Site 2: _____________mm

   Site 3: _____________mm

   Calculation of Body Fat Percentage: ____________________________

   Classification: ____

   *Note: You may include more sites at your own discretion.

3. **Body Mass Index (BMI)**

   Date of Test: _________________

   Raw Score: _________________

   Height: ________m Weight: _______kg

   BMI = ______________ Classification: _________________________

4. Waist to Hip Ratio

Date of Test: ________________

Waist: ____________cm Hip: ____________cm

Ratio: Waist/hip = _______________ Classification: _____________

5. Bioelectrical Impedance Analysis (BIA)

Date of Test: ____________________

Calculation of Body Fat Percentage: ___________________

Classification: _____________________________________

Other results from BIA: _____________________________________

Questions (40 marks)

1. Define lean body weight. (3 marks)
2. Identify potential sources of error in determining body composition using skinfold analysis. (5 marks)
3. Tabulate the results of the various measurements of body composition assessment on yourself. Compare and contrast the various methods. (22 marks)
4. Based on your results from the DEXA, comment on your bone density. (10 marks)

Laboratory Session 2

Weight Management

Maintaining a healthy weight could be an indicator of health status.

Table 1. Calculations for desirable body weight (adapted from Howley & Franks, 1986).

Desirable Body Weight
Fat weight = current weight × (% fat/100)
Lean body weight (LBW) = current weight − fat weight
Desirable weight = LBW/ 1 − (% fat desired/100)
Desirable fat loss = current weight − desirable body weight
For example, assume a man weighs 190 lb and his body fat is 22%. 15% is a desirable body fat percentage for men (1 lb = 0.45 kg).
Fat weight = 190 lb × 22/100 = 41.8 lb
LBW = 190 − 41.8 = 148.2 lb
Desirable weight = 148.2/(1 − (15%/100)) = 174.4 lb
Desirable fat loss = 190 – 174.4 lb = 15.6 lb

Many training or exercise programmes are slowly geared towards modifying body size and composition, i.e. reduce size of fat cells and increase muscle mass (Balasekaran, Loh, Govindaswamy, & Cai, 2014). Preventing the increase in fat mass can be achieved through the basic principle of eating in moderation and ensuring total caloric intake is equivalent or less than energy output (Table 1, Figures 1, 2, & 3).

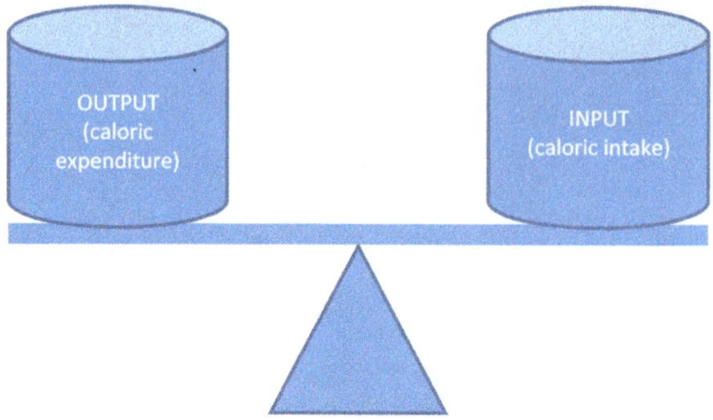

Figure 1. Equivalent energy intake and output resulting in no fat mass loss.

Figure 2. Greater caloric intake than output results in greater stores of fat mass.

Figure 3. Greater caloric expenditure than caloric intake results in lower stores of fat mass.

There are desirable body weights based on different populations. Below is a chart that shows desirable body weight for both female and males for ages 25 and over (Tables 2 & 3). These tables were introduced by insurance agents so that they could assess their client's health risk if they

Table 2. Desirable weight for females age 25 and above (adapted from Metropolitan Life Insurance Company, 1983).

Height		Small Frame		Medium Frame		Large Frame	
in.	cm	lb	kg	lb	kg	lb	kg
57	145	99–108	45–49	106–118	48–54	115–128	52–58
58	147	100–110	45–50	108–120	49–55	117–131	53–60
59	150	101–112	46–51	110–123	50–56	119–134	54–61
60	152	103–115	47–52	112–126	51–57	122–137	55–62
61	155	105–118	48–54	115–129	52–59	125–140	57–64
62	157	108–121	49–55	118–132	54–60	128–144	58–65
63	160	111–124	50–56	121–135	55–61	131–148	60–67
64	163	114–127	52–58	124–138	56–63	134–152	61–69
65	165	117–130	53–59	127–141	58–64	137–156	62–71
66	168	120–133	55–60	130–144	59–65	140–160	64–73
67	170	123–136	56–62	133–147	60–67	143–164	65–75
68	173	126–139	57–63	136–150	62–68	146–167	66–76
69	175	129–142	59–65	139–153	63–70	149–170	68–77
70	178	132–145	60–66	142–156	65–71	152–173	69–79
71	180	135–148	61–67	145–159	66–72	155–176	70–80

*Height measured without shoes.

were buying insurance policies. It is an easy way to determine health risk and it is only available for adults as they are in need for insurance policies. Health professionals can use these tables in mass exercise settings so that it provides the assessor an idea of their client's health risk or develop a goal for fitness before they embark on an exercise programme.

Attaining a desired body weight based on height is a simple tool for the general population, but some simple calculations can help individuals understand the percentage of their fat mass and the recommended fat percentage for different age groups. Health professionals can scan and choose the various methods from Laboratory Session 1 to help their clients understand body fat measurement as the chapter provides a deeper understanding of the various methods to measure fat mass and lean mass.

Table 3. Desirable weight for males age 25 and above (adapted from Metropolitan Life Insurance Company, 1983).

Height		Small Frame		Medium Frame		Large Frame	
in.	cm	lb	kg	lb	kg	lb	kg
61	155	123–129	56–59	126–136	57–62	133–145	60–66
62	157	125–131	57–60	128–138	58–63	135–148	61–67
63	160	127–133	58–60	130–140	59–64	137–151	62–69
64	163	129–135	59–61	132–143	60–65	139–155	63–70
65	165	131–137	60–62	134–146	61–66	141–159	64–72
66	168	133–140	60–64	137–149	62–68	144–163	65–74
67	170	135–143	61–65	140–152	64–69	147–167	67–76
68	173	137–146	62–66	143–155	65–70	150–171	68–78
69	175	139–149	63–68	146–158	66–72	153–175	70–80
70	178	141–152	64–69	149–161	68–73	156–179	71–81
71	180	144–155	65–70	152–165	69–75	159–183	72–83
72	183	147–159	67–72	155–169	70–77	163–187	74–85
73	185	150–163	68–74	159–173	72–79	167–192	76–87
74	188	153–167	70–76	162–177	74–80	171–197	78–90
75	191	157–171	71–78	166–182	75–83	176–202	80–92

*Height measured without shoes.

3-Day Activity and Dietary Recall*

Caloric measurement is the basis of quantifying calories to determine daily caloric gain or deficit. For weight management, caloric deficit is of interest in this chapter. The American College of Sports Medicine (ACSM) recommends a caloric deficit of 500 kcal per day through exercise to promote healthy weight loss, which can be monitored via the activity and dietary recall activity (Figures 1, 2, & 3). Caloric deficit is achieved when total caloric output (exercise and daily activities) is more than caloric intake. Therefore, the total calories consumed daily is less than the calories expended for day's activities.

The solution to monitor caloric intake and expenditure could be the use of a dietary and activity log sheet, respectively. Both log sheets facilitate the recall and monitoring of caloric input and output with reference to the individual's food intake habits through the day.

*A minimum of 1,200 kcal for women and 1,800 kcal for men is necessary for daily human body function. A maximum of 0.9 kg weight loss per week for any weight loss programme is recommended (ACSM, 2018).

24-Hour Nutrition Log Sheet

To calculate calories consumed daily, an individual can adapt the dietary log sheet listed in McArdle, Katch, and Katch (2007) or a similar dietary log sheet (Figure 5). For each food type, the quantity and weight of each macronutrient must be listed. A reference sheet for different food sources can also be identified from McArdle, Katch, and Katch (2007) or from respective country's health ministries (e.g. Singapore Health Promotion Board, Singapore Ministry of Health (Figure 4)). The representation of the total calorie intake by nutrient type (carbohydrate, protein, and fats) helps individuals understand their nutrient contribution from their food intake. The proportion of caloric content per nutrient per food item should only be taken as reference as caloric content varies across food manufacturers. For compound food items such as chicken rice, which is not listed in the table, it can be deconstructed as 'rice' and 'chicken'. Table 4 shows an example of a completed 24-hour dietary recall log sheet.

The calorie per gram for both carbohydrate and protein are 4 kcal each, and fats are 9 kcal. To calculate the caloric intake per nutrient type, multiply the total grams of each nutrient by its kcal/gram. To calculate the total calorie intake, add the total calorie of each nutrient type (Table 5). The recommended daily diet ratio intake of macronutrients for carbohydrates: 50–60%, protein: 10–15%, fat: 25–30%.

Food Serving Guideline

Food Group	Number of Servings per Day	Example of 1 Serving	Remarks
Rice and Alternatives	**5 to 7**	2 slices bread (60 g) ½ bowl rice (100 g) ½ bowl noodles or beehoon (100 g) 4 plain biscuits (40 g) 1 thosai (60 g) 2 small chapatis (60 g) 1 large potato (180 g) 1½ cup plain cornflakes (40 g)	
Fruits	**2**	1 small apple, orange, pear, or mango (130 g) 1 wedge pineapple, papaya, or watermelon (130 g) 10 grapes or longans (50 g) 1 medium banana ¼ cup dried fruit (40 g) 1 glass pure fruit juice (250ml)	1 glass – Half a bottle of Ice Mountain mineral water
Vegetables	**2**	¾ mug cooked leafy or non-leafy vegetables (100 g) ¼ round plate of cooked vegetables 150 g raw leafy vegetables 100 g raw non-leafy vegetables	1 mug – Half a bottle of Ice Mountain mineral water
Meat and Alternatives	**2 to 3**	1 palm-sized piece fish, lean meat, or skinless poultry (90 g) 2 small blocks soft beancurd (170 g)	~ half of 
		¾ cup cooked pulses (e.g. lentils, peas, beans) (120 g)	
		5 medium prawns (90 g) 3 eggs (150g)++	Egg yolks are high in cholesterol, eat no more than 4 egg yolks a week.
		2 glasses of milk (500 ml) 2 slices of cheese (40 g)	2 glasses – volume of a bottle of Ice Mountain mineral water i.e. 1 slice – 1 slice of Chesdale cheese

Figure 4. Food serving guideline (adapted from the Singapore Ministry of Health).

My Food Diary (WK 1: 11 Jan - 17 Jan 2021)						
Day & Date	**Meal**	**Time**	**Food**	**Quantity**	**Method of Cooking**	**Remarks**
Monday, 11-01-2021	Breakfast					
	Meal	**Time**	**Food**	**Quantity**	**Method of Cooking**	**Remarks**
	Lunch					
	Meal	**Time**	**Food**	**Quantity**	**Method of Cooking**	**Remarks**
	Dinner					
	Meal	**Time**	**Food**	**Quantity**	**Method of Cooking**	**Remarks**
	Others (i.e. Tea break, supper)					

Figure 5. Sample of a weekly food diary.

Day 1 (repeat the same for Days 2 and 3) (Table 5)

Total grams of carbohydrate = ________ × 4 kcal/gram

Total kcals from carbohydrate = ________ kcal

Total grams of fat = ________ × 9 kcal/gram

Total kcals from fat = ________ kcal

Total grams of protein = ________ × 4 kcal/gram

Total kcals from protein = ________ kcal

Day 1: Total kcals = ________

Day 2: Total kcals = ________

Day 3: Total kcals = ________

Table 4. 24-hour diet recall log sheet.

Food	Quantity	Carbohydrate (grams)	Protein (grams)	Fat (grams)
Breakfast				
Egg sandwich	2 slices	14.5	4.82	9.8
Instant coffee	160 g	4.8	0.256	0
Plain water	1 glass	0	0	0
Lunch				
Tuna sandwich (with tomato and lettuce)	2 slices	12.82	8.14	4.84
Tomato		1.23	0.251	0.060
Lettuce		1.18	0.572	0.108
Whole milk	120 ml	15.84	11.16	11.4
Dinner				
Cooked rice	100 g	68.6	0.567	0.3
Roasted chicken	40 g	0	35.08	5.12
Plain water	2 glasses	0	0	0
Whole milk	120 ml	15.84	11.16	11.4
Supper				
Hot cocoa with milk	160 g	46.4	16.48	0
Turkey sandwich (with tomato and lettuce)	2 slices	12.82	8.14	4.84
Tomato		1.23	0.251	0.060
Lettuce		1.18	0.572	0.108
	Total:	196.44 g	97.45 g	48.04 g

Table 5. Calculation of total calorie intake by nutrient type.

Macronutrient	Total grams	kcal/gram	Total kcal (total grams × kcal/ gram)
Carbohydrates	196.44	4	785.76
Protein	97.45	4	389.80
Fats	48.04	9	432.36
		Total:	1607.92 kcal

Average kcals per day: (Day 1 + Day 2 + Day 3)/3 = ________ kcals

Day 1: Total kcals from carbohydrate/total kcals x 100% = ________ % carbohydrates

Total kcals from fat/total kcals x 100% = ________ % fat

Total kcals from protein/total kcals x 100% = ________ % protein

Day 2: Total kcals from carbohydrate/total kcals x 100% = ________ % carbohydrates

Total kcals from fat/total kcals x 100% = ________ % fat

Total kcals from protein/total kcals x 100% = ________ % protein

Day 3: Total kcals from carbohydrate/total kcals x 100% = ________ % carbohydrates

Total kcals from fat/total kcals x 100% = ________ % fat

Total kcals from protein/total kcals x 100% = ________ % protein

3 days average:

kcals from carbohydrates on Day 1 + Day 2 + Day 3/3 days total kcals x 100% = ________ % carbohydrates

kcals from fat on Day 1 + Day 2 + Day 3/3 days total kcals x 100% = ________ % fat

kcals from protein on Day 1 + Day 2 + Day 3/3 days total kcals x 100% = ________ % protein

24-Hour Activity Log Sheet

Caloric expenditure per activity is dependent on weight, duration, and intensity. Hence, caloric output per individual is different even when assigned identical intensity-level activities. To monitor daily caloric expenditure, it is ideal to record the activities carried out through the day. For each activity, the calorie per minute can be identified from the reference chart (McArdle, Katch, & Katch, 2007). First, identify the activity closest to the activity performed from the reference chart. Next, identify the calorie per minute according to the weight closest to the current body mass. An example of a completed 24-hour activity log sheet for an individual of 62 kg (Table 6) and a weekly physical activity log sheet (Figure 6) are shown on the next page.

My Personal Physical Activity Log
(WK 1: 11 Jan - 17 Jan 2021)

*Optional: refer to the OMNI RPE scale (Robertson, 2004)

Day & Date	Description of Activity	Duration (hr/min)	(RPE 0 to 10)	Remarks
Monday, 11-01-2021				
Day & Date	**Description of Activity**	**Duration (hr/min)**	**(RPE 0 to 10)**	**Remarks**
Tuesday, 12-01-2021				
Day & Date	**Description of Activity**	**Duration (hr/min)**	**(RPE 0 to 10)**	**Remarks**
Wednesday, 13-01-2021				
Day & Date	**Description of Activity**	**Duration (hr/min)**	**(RPE 0 to 10)**	**Remarks**
Thursday, 14-01-2021				
Day & Date	**Description of Activity**	**Duration (hr/min)**	**(RPE 0 to 10)**	**Remarks**
Friday, 15-01-2021				
Day & Date	**Description of Activity**	**Duration (hr/min)**	**(RPE 0 to 10)**	**Remarks**
Saturday, 16-01-2021				
Day & Date	**Description of Activity**	**Duration (hr/min)**	**(RPE 0 to 10)**	**Remarks**
Sunday, 17-01-2021				

Figure 6. Sample of a weekly physical activity log sheet. The Children OMNI Rate of Perceived Exertion (RPE) scale (Robertson et al., 2000; Utter et al., 2004) and Adult OMNI RPE scale (Robertson et al., 2000) can also be used to monitor your exercise intensity.

Table 6. 24-hour activity log sheet (an example of an elite athlete. Note: energy expenditure is very high).

Activity	Time (hh:mm)	Duration (min)	kcal/min	Total energy cost (kcal)
Wake up	9:00 am			
Morning run	9:30 am	30	14.2	426
Shower	9:45 am	15	1.4	21
Breakfast	10:45 am	60	1.6	96
Walk to classroom	11:00 am	15	5.4	81
Complete assignment on computer	2:00 pm	180	1.4	252
Lunch	2:45 pm	45	1.6	72
Walk home	3:00 pm	15	5.4	81
Afternoon nap	4:00 pm	60	1.4	84
Read	6:00 pm	120	1.4	168
Warm-up	6:15 pm	15	5.1	76.5
Run	7:15 pm	60	14.2	852
Strength workout	8:00 pm	15	5.1	229.5
Shower	8:30 pm	30	1.4	42
Dinner	9:30 pm	60	1.6	96
Read	10:30 pm	60	1.4	84
Listen to music	11:30 pm	60	1.4	84
Sleep	12:30 am	570	1.4	798
	Total:	1,440 mins	Total daily energy cost:	3,543 kcal

Alternatively, total energy expenditure (TEE) can be calculated by using the following equation (Table, 2005):

TEE = A + B × age + PA × (D × weight + E × height)

A = Constant term
PA = Physical activity coefficient[1]
B = Age coefficient
D = Weight coefficient
E = Height coefficient

[1]PA Category
Sedentary: 1.0–1.39
Low active: 1.4–1.59
Active: 1.6–1.89
Very active: 1.9–2.49

For males:

TEE = 864 – 9.72 × age (years) + PA × [(14.2 × weight (kg) + 503 × height (metres)]

For females:

TEE = 387 – 7.31 × age (years) + PA × [(10.9 × weight (kg) + 660.7 × height (metres)]

(Gerrior, Juan, & Peter, 2006)

Both recall logs help to keep track of and monitor caloric intake and expenditure with reference to the food consumed and activity for the specific day, respectively.

Taking into consideration day-to-day variations in activity and food consumed, a 3-day dietary and activity log is recommended. Hence, the above examples of dietary and activity log sheets should be repeated twice (include one weekend day). Tables 7, 8, 9, and 10 show samples of the total calculations of an individual's daily and weekly caloric input and output. Below is a summary of the steps to calculate the average of 3 days:

Step 1: Complete a dietary log and activity log sheet daily
Step 2: Calculate total daily calorie and total daily energy cost
Step 3: Add Days 1, 2, and 3 of total daily calorie intake then divide by 3
Step 4: Add Days 1, 2, and 3 of total daily energy cost then divide by 3

Table 7. Approximate daily caloric input and output (an example of an elite athlete. Note: energy expenditure is very high).

Daily Caloric requirement	Daily Caloric Intake	Caloric Balance	Caloric Output (Body function & Physical Activities)
Average: 2,000 kcal Minimum: 1,200 kcal* 1,800 kcal*	1,607.92 kcal	−392.08 kcal	3,543 kcal

*Minimum 1,200 kcal for females, 1,800 kcal for males

Table 8. Five day caloric intake and output (an example of an elite athlete. Note: energy expenditure is very high).

Caloric requirement	Caloric Intake	Caloric Balance	Caloric Output (Body function & Physical Activities)
10,000 kcal	8,039.6 kcal	−1,960.4 kcal	17,715 kcal

Table 9. Daily caloric input versus output (an example of an elite athlete. Note: energy expenditure is very high).

Caloric Intake Balance: −392.08 kcal	**Energy Expenditure: 3,543 kcal**	✓ **Good**	

Table 10. Current diet and exercise status (note: for this elite athlete, the energy expenditure is very high and she meets the minimum criteria for caloric intake. However, increasing her food intake will be beneficial as she needs more energy to perform well in her sport).

Caloric Intake (Food)	 **Low (Food Intake)**	**Caloric Output (Body Function and Physical Activities)**	 **High Energy Expenditure**	✓ **Good**	

Questions (40 marks)

1. Comment on your desired body weight using your height and weight from Tables 2 or 3. (5 marks)
2. Calculate your fat weight and lean body weight (LBW). Comment on your health status (Table 1). (10 marks)
3. Calculate your desirable weight using the formula enclosed above in Table 1. Based on the results, what kind of exercise or diet recommendations would you advocate? (15 marks)
4. Why does a skinfold technique give us a better picture of body composition than height and weight tables (Tables 2 or 3)? (10 marks)

Laboratory Session 3

Resting Metabolic Rate — Chicken Laboratory

Our body constantly uses energy to keep the systems functioning regardless of waking or resting state. The process of consumed energy and heat generated is termed as energy metabolism. The rate at which energy is consumed is measured through calories (kcal) via direct or indirect calorimetry (Figure 1).

Measurement of direct calorimetry through heat production requires long hours and cannot be applied in real-time measure of sports and most physical activities. Indirect calorimetry measures the rate of oxygen (O_2) consumption and carbon dioxide (CO_2) produced associated with metabolising nutrients and requires a smaller window of fasting. Indirect calorimetry measurement is the gold standard of caloric measurement due to its accurate, valid, and convenient measure of energy consumed (Figure 2).

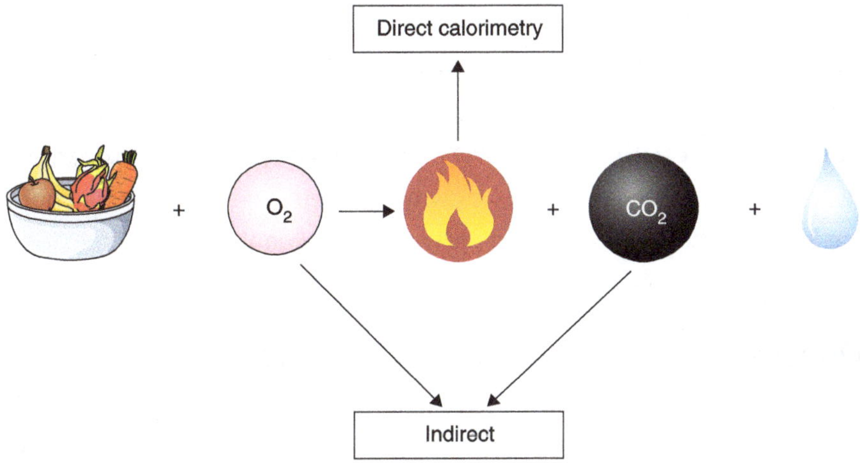

Figure 1. Mode of measurement of direct and indirect calorimetry.

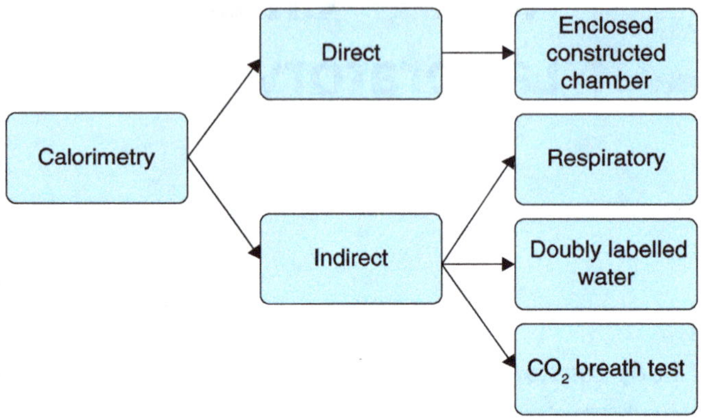

Figure 2. Types of direct and indirect calorimetry measurement.

In addition, technology advancement has improved accuracy in the tracking and measurement of the rate of metabolism. In clinical settings, indirect calorimetry is the standard technique in the clinical armamentarium against which all other methods of nutritional assessment should be measured (Wilson, Grande, & Hoyt, 2007).

Calorimetry measurements can measure both basal metabolic rate (BMR) and resting metabolic rate (RMR). BMR refers to the minimum amount of energy required to keep the body systems functioning, typically measured in a laboratory or using formula. However, the laboratory protocol would require 14–18 hours of fasting for a 15–20-minute measurement. In comparison, RMR can be tested with a 4-hour fasting period. RMR is defined as the daily caloric expenditure in a relaxed, awake, and resting state. Similar for both BMR and RMR, they are affected by various factors such as genes, lifestyle, physical activity, diet, etc. (Table 1). Do also note that some authors use BMR and RMR interchangeably.

Measurement of BMR (Non-Laboratory)

The relationship between heat release and body mass is long understood as smaller animals have a greater heat production rate than larger animals. The unit of measurement for BMR is expressed as $kcal{\cdot}m^{-2}{\cdot}h^{-1}$.

Table 1. Influence of various factors and resting metabolic rate (RMR).

Variables	Influence	Explanation
Genes		Higher correlation in monozygotic than in dizygotic twins whether expressed per kg of body weight or muscle mass
Age (Fukagawa, Bandini, & Young, 1990; Melzer, 2011)	Decrease	RMR decreases with age independent of decrease in muscle mass (Tables 2 & 3, Figure 3)
Gender	M > F	RMR is higher in men than in women, independent of differences in body composition, fitness level, menopausal status, or age
Growth	Higher during growth	RMR is increased during growth due to the synthesis of growing tissues and the energy deposited in those tissues
Muscle mass	High	Higher RMR is observed in highly-trained individuals as compared to moderately-trained and untrained individuals
Exercise (Poehlman, 1989)	Increase	Purposeful physical exercise increases energy expenditure. Highly-trained individuals have the highest RMR as compared to moderately-trained and untrained individuals (Table 4)
Illness/Stress	Increase	Injury, fever, surgery, renal failure, burns, and infection increase RMR
Pregnancy	Increase	RMR increases in late pregnancy by approximately 20% mainly due to increase of maternal body mass

Table 2. Participants' descriptive statistics.

	Group			P Value	
	Young Men (n = 24)	Old Men (n = 24)	Old Women (n = 20)	Young versus Old Men	Old Men versus Women
Age (year)	21 ± 1	75 ± 1	72 ± 1	0.001	0.05
Height (cm)	176.6 ± 1.7	172.1 ± 1.2	158.8 ± 1.2	0.05	0.001
Weight (kg)	72.3 ± 2.1	72.8 ± 1.6	59.4 ± 2.2	0.41	0.001
Body mass index ($kg{\cdot}m^{-2}$)	23.1 ± 0.5	24.6 ± 0.5	23.5 ± 0.8	0.05	0.25
Fat-free mass* (kg)	55.4 ± 1.9	47.7 ± 0.9	35.8 ± 0.8	0.001	0.001
Total body fat* (kg)	16.9 ± 1.0	25.1 ± 1.0	23.6 ± 1.5	0.001	0.41
% Body fat*	23 ± 1	34 ± 1	39 ± 1	0.001	0.003

Values are mean ± standard deviation.

*Estimated by isotope dilution (adapted from Fukagawa, Bandini, & Young, 1990; Melzer, 2011).

Table 3. Metabolic rate in young men, old men, and old women.

Resting Metabolic Rate	Young Men	Old Men	Old Women
Mean Measured	1.24 ± 0.03	1.04 ± 0.02*	0.84 ± 0.02**
Adjusted for FFM	1.13 ± 0.02	1.03 ± 0.02*	0.99 ± 0.02***
WHO/FAO/UNU Estimate ****	1.23 ± 0.02	1.02 ± 0.01	0.85 ± 0.02

Values are mean ± standard deviation in kcal•min^{-1}. Fat-free mass (FFM).

*$p < 0.002$ for comparison with young men.

**$p < 0.001$ for comparison with old men.

***$p = 0.16$ for comparison with old men (adapted from Fukagawa, Bandini, & Young, 1990; Melzer, 2011).

****See references for estimation (Fukagawa, Bandini, & Young, 1990).

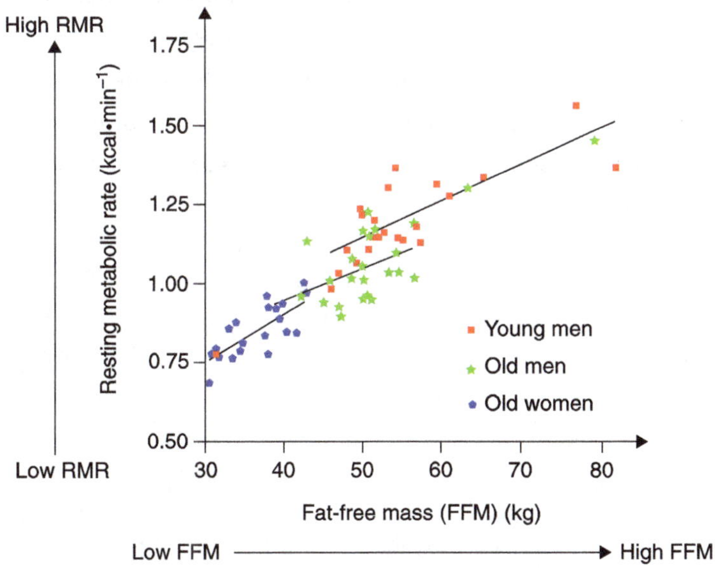

Figure 3. Relationship between resting metabolic rate and fat-free mass estimated by isotope dilution in young men, old men, and old women. Overall, $r = 0.89$, $p < 0.001$ (adapted from Fukagawa, Bandini, & Young, 1990; Melzer, 2011).

Table 4. Exercise affects resting metabolic rate (RMR) (adapted from Poehlam et al., 1989).

	Untrained (UT) (n = 9)	Moderately-trained (MT) (n = 11)	Highly-trained (HT) (n = 8)
RMR (kcal•min^{-1})	1.24	1.13	1.31*
RMR (kcal•kg^{-1}hr^{-1})	0.93	0.92	1.11*
RMR (kcal•FFW^{-1}hr^{-1})	1.09	1.06	1.20*

*Indicates significant difference between groups:

RMR (kcal•min^{-1}) (HT versus MT)

RMR (kcal•kg^{-1}hr^{-1}) (HT versus UT; HT versus MT)

RMR (kcal•FFW^{-1}hr^{-1}) (HT versus UT; HT versus MT).

The relationship between heat production and body mass had been established by research conducted by various researchers. Interventions conducted on animals revealed consistent range of heat release and smaller animals, such as rats, have greater heat production than larger animals like elephants. Figure 4 below shows a positive linear correlation between BMR and body mass.

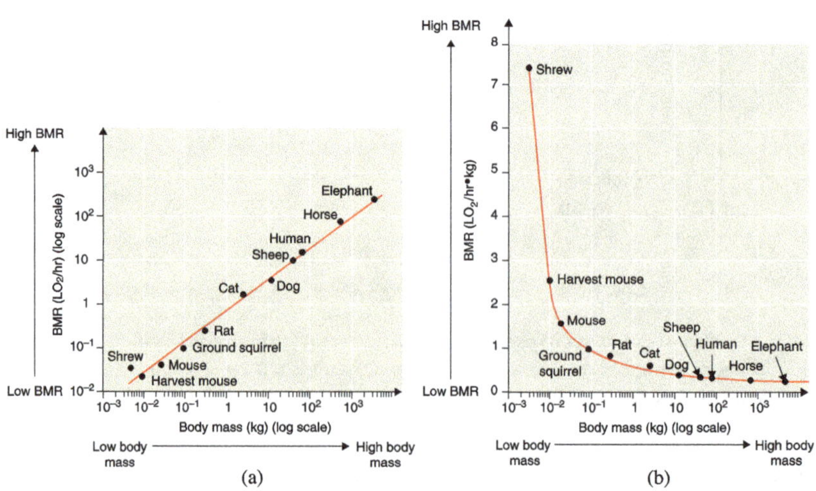

Figure 4. a) Correlation between BMR and body sizes of various mammals. b) Correlation between BMR per kilogram of body mass and body size (adapted from McArdle, Katch, & Katch, 2010) (note the link between human body mass and BMR).

A case study of Method 1 and Method 2 are described below:

1) Method 1: Harris-Benedict equation (Table 5)
2) Method 2: Use of nomogram and standard basal metabolic rate chart (Figure 5)

Table 5. Harris-Benedict equation for men and women.

Men	Women
BMR = 66.5 + (5 × H) + (13.7 × W) − (6.8 × A)	BMR = 66.5 + (1.9 × H) + (9.5 × W) − (4.7 × A)

H refers to height expressed in centimetres (cm); W refers to weight expressed in kilograms (kg); A refers to age expressed in years.

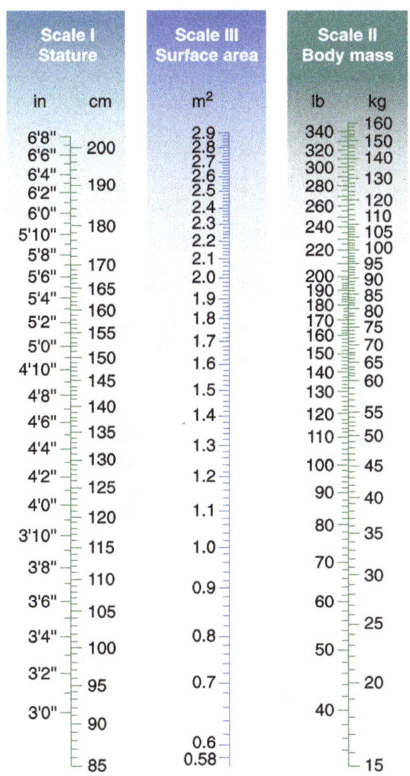

Figure 5. Nomogram to estimate body surface area from stature and mass (adapted from Dubois, 1936; McArdle, Katch, & Katch, 2010).

Table 6. Standard basal metabolic rates (adapted from Fleish, 1951; McArdle, Katch, & Katch, 2010).

	$kcal \cdot m^{-2} \cdot h^{-1}$		
Age (years)	Men	Women	
1	53.0	53.0	High ↑
2	52.4	52.4	
3	51.3	51.2	
4	50.3	49.8	
5	49.3	48.4	
6	48.3	47.0	
7	47.3	45.4	
8	46.3	43.8	
9	45.2	42.8	
10	44.0	42.5	
11	43.0	42.0	
12	42.5	41.3	
13	42.3	40.3	
14	42.1	39.2	
15	41.8	37.9	
16	41.4	36.9	
17	40.8	36.3	
18	40.0	35.9	
19	39.2	35.5	
20	38.6	35.3	
25	37.5	35.2	
30	36.8	35.1	
35	36.5	35.0	
40	36.3	34.9	
45	36.2	34.5	
50	35.8	33.9	
55	35.4	33.3	
60	34.9	32.7	
65	34.4	32.2	
70	33.8	31.7	
75+	33.2	31.3	↓ Low

	$kj \cdot m^{-2} \cdot h^{-1}$		
Age (years)	Men	Women	
1	222	222	High ↑
2	219	219	
3	215	214	
4	211	208	
5	206	203	
6	202	197	
7	198	190	
8	194	183	
9	189	179	
10	184	178	
11	180	176	
12	178	173	
13	177	169	
14	176	164	
15	175	159	
16	173	154	
17	171	152	
18	167	150	
19	164	149	
20	162	148	
25	157	147	
30	154	147	
35	153	146	
40	152	146	
45	152	144	
50	150	142	
55	148	139	
60	146	137	
65	144	135	
70	141	133	
75+	139	131	↓ Low

Adapted from Fleish A. (1951). Le metabolisme basal standard et sa determination au moyen du "Metabocalculator." *Helv Med Acta, 18*(1), 23–44.

Note: The nomogram determines the surface area of the body, following which multiply it according to the standard basal metabolic rate (Table 6), which is ± 10% of standard normal BMR (McArdle, Katch, & Katch, 2010).

Daily caloric needs by appropriate activity factor:

- Sedentary (little or no exercise) = BMR × 1.2
- Lightly active (light exercise/sports 1–3 days/week) = BMR × 1.375
- Moderately active (moderate exercise/sports 3–5 days/week) = BMR × 1.55
- Very active (hard exercise/sports 6–7 days/week) = BMR × 1.725
- Extra active (very hard exercise/sports and physical job) = BMR × 1.9

Case Study

Carla is a sedentary 19 years old. She measures 170 cm tall and weighs 60 kg. Find her BMR.

Method 1: Harris-Benedict equation (Table 5)

$$\begin{aligned} \text{BMR} &= 66.5 + (1.9 \times \text{H}) + (9.5 \times \text{W}) - (4.7 \times \text{A}) \\ &= 66.5 + (1.9 \times 170) + (9.5 \times 60) - (4.7 \times 19) \\ &= 66.5 + 323 + 570 - 89.3 \\ &= 870.2 \text{ kcal} \end{aligned}$$

As a sedentary individual, multiply kcal/day by 1.2. Therefore, Carla's daily caloric needs would be: 870.2 × 1.2 = 1044.24 kcal

Method 2: Use of nomogram and standard basal metabolic rate chart

With reference to the nomogram (Figure 5), her body surface area is 1.76 m^2.

According to the standard basal metabolic rate chart (Table 6), her metabolic rate at 19 years old is 35.5 $kcal \cdot m^{-2} \cdot hr^{-1}$.

Multiply 35.5 $kcal \cdot m^{-2} \cdot hr^{-1}$ with 1.76 m^2 = 62.48 $kcal \cdot hr^{-1}$

Her BMR is 62.48 $kcal \cdot hr^{-1}$.

For 24 hours, her daily caloric expenditure is 62.48 $kcal \cdot hr^{-1}$ × 24 hrs = 1499.52 kcals

Therefore, daily caloric expenditure would be 1499.52 kcal.

Chicken Laboratory Protocol

The purpose of the laboratory test is to evaluate RMR during the post-absorptive state immediately from a high-protein meal. Energy used during the post-prandial state for rest or physical activity is derived from carbohydrates and/or fats. The type of fuel catabolized for energy can also be determined by the respiratory exchange ratio (RER) which is a division of volume of carbon dioxide over volume of oxygen (VCO_2/VO_2). Tables 7 and 8 indicate caloric values of oxygen for non-protein ratios.

Procedure (Table 9 & Figure 6):

1) Calibrate metabolic cart and prepare equipment (canopy head and heart rate monitor)
2) Allow participant to rest for 10–15 minutes
3) Pre-measurement test: put participant under indirect calorimetry metabolic cart (dilution method with canopy head) for 30 minutes
4) Provide high-protein diet meal (i.e. whole chicken) for participant
5) Let participant rest for 30 minutes after consumption of high-protein diet meal
6) Post-measurement test: put participant under indirect calorimetry metabolic cart (dilution method with canopy head) for 30 minutes

Table 7. Caloric equivalence of the RER and % kcal from carbohydrates and fats (adapted from Wilmore & Costill, 1999).

	Energy	% kcal	
RER	kcal/LO_2	Carbohydrates	Fats
0.71	4.69	0.0	100.0
0.75	4.74	15.6	84.4
0.80	4.80	33.4	66.6
0.85	4.86	50.7	49.3
0.90	4.92	67.5	32.5
0.95	4.99	84.0	16.0
1.00	5.05	100.0	0.0

Table 8. Caloric values of oxygen for non-protein exchange ratios (adapted from Zuntz, 1901; McArdle, Katch, & Katch, 2010).

RER	kcal/L	RER	kcal/L	RER	kcal/L
.70	4.686	.80	4.801	.90	4.924
.71	4.690	.81	4.813	.91	4.936
.72	4.702	.82	4.825	.92	4.948
.73	4.714	.83	4.838	.93	4.960
.74	4.727	.84	4.850	.94	4.973
.75	4.739	.85	4.863	.95	4.985
.76	4.752	.86	4.875	.96	4.997
.77	4.764	.87	4.887	.97	5.010
.78	4.776	.88	4.900	.98	5.022
.79	4.789	.89	4.912	.99	5.034
				1.00	5.047

*Values are in respiratory exchange ratio (RER) and kilocalories per litre (kcal/L).

Note: More detailed version.

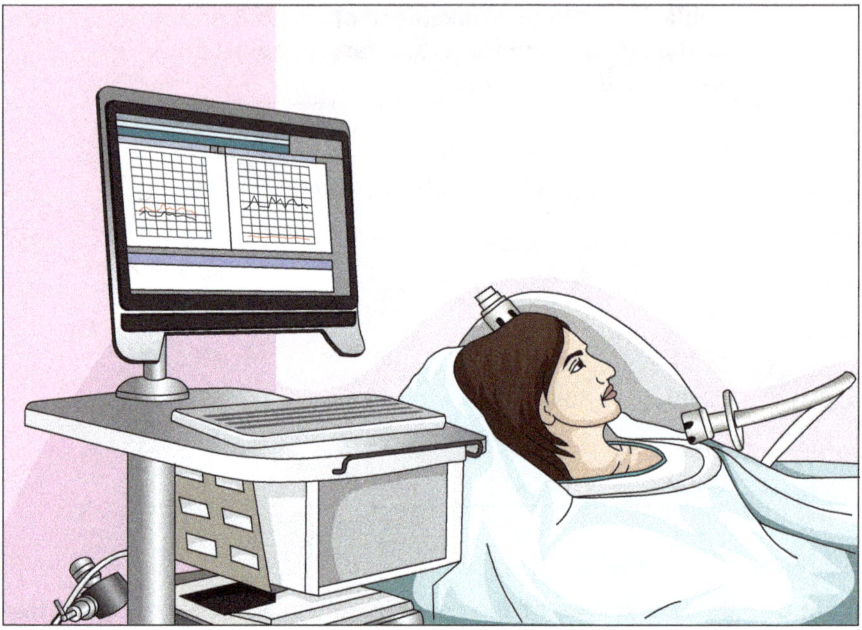

Figure 6. Indirect calorimetry metabolic cart (dilution method with canopy head).

Table 9. Definition of terms.

V_E	Amount of air expired per minute
VO_2	Amount of oxygen consumed per minute
VCO_2	Amount of carbon dioxide expired per minute
RER	Respiratory exchange ratio — ratio of volume of CO_2 expired per minute (VCO_2) to the volume of O_2 consumed (VO_2) during the same time interval (i.e. VCO_2/VO_2)
Relative RER	(RMR – NBMR)/NBMR (refer to Table 6 for norm basal metabolic rate)

Resting Metabolic Rate (RMR) Calculation for Participant

Step 1: Calculate pre-meal RMR values

Average VO_2 (AVO_2) in 30 mins = ________ $L \cdot min^{-1}$

Average VCO_2 ($AVCO_2$) in 30 mins = _______ $L \cdot min^{-1}$

RER = $AVCO_2/AVO_2$

= _______

kcal/L based on RER (Refer to Table 8) = _______

Use nomogram (Figure 5) to estimate body surface area (BSA).

OR

Use one of the following formulas to calculate BSA:

1. DuBois & DuBois (1916): 0.20247 x height $(m)^{0.725}$ x weight $(kg)^{0.425}$
2. Haycock, Schwartz, & Wisotsky (1978) : 0.024265 x height $(cm)^{0.3964}$ x weight $(kg)^{0.5378}$
3. Gehan & George (1970): 0.0235 x height $(cm)^{0.42246}$ x weight $(kg)^{0.51456}$
4. Boyd (1935): 0.0003207 x height $(cm)^{0.3}$ x weight $(grams)^{0.7285 - (0.0188 \times \log(weight))}$

BSA = _______ m^2

RMR = (AVO_2 x RER* x 60)/BSA

= __________ $kcal \cdot m^{-2} \cdot hr^{-1}$

*kcal/L

Norm Basal Metabolic Rate (NBMR): based on participant's age (Table 6)

Relative RMR change (%) = (RMR – NBMR)/NBMR x 100% = ________%

Daily Energy Expenditure of Participant

RMR: __________ $kcal \cdot m^{-2} \cdot hr^{-1}$

BSA: ____________ m^2

RMR ($kcal \cdot m^{-2} \cdot hr^{-1}$) x BSA ($m^2$) x 24 hours = ________________ kcal/day

Step 2. Calculate post-meal RMR values using Step 1 (after consumption of chicken).

Average VO_2 (AVO_2) in 30 mins = ________ L•min^{-1}

Average VCO_2 ($AVCO_2$) in 30 mins = _______ L•min^{-1}

RER = VCO_2/VO_2

= _______

kcal/L based on RER (Refer to Table 8) = _______

Use nomogram (Figure 5) to estimate BSA.

OR

Use the following formulas to calculate BSA:

1. DuBois & DuBois (1916): 0.20247 x height (m)$^{0.725}$ x weight (kg)$^{0.425}$
2. Haycock, Schwartz, & Wisotsky (1978) : 0.024265 x height (cm)$^{0.3964}$ x weight (kg)$^{0.5378}$
3. Gehan & George (1970): 0.0235 x height (cm)$^{0.42246}$ x weight (kg)$^{0.51456}$
4. Boyd (1935): 0.0003207 x height (cm)$^{0.3}$ x weight (grams)$^{0.7285-(0.0188 \times \log(\text{weight}))}$

BSA = _______ m^2

RMR = (AVO_2 x RER* x 60)/BSA

= ___________ kcal•m^{-2}•hr^{-1}

* kcal/L

Norm Basal Metabolic Rate (NBMR): based on participant's age (Table 6)

Relative RMR change (%) = (RMR – NBMR)/NBMR x 100% = ________%

Daily Energy Expenditure of Participant

RMR: ___________ kcal•m^{-2}•hr^{-1}

BSA: ___________m^2

RMR (kcal•m^{-2}•hr^{-1}) x BSA (m^2) x 24 hours = ________________ kcal/day

Step 3. Report the relative change in RMR as a result of the protein meal (%).

RMR (pre-meal): ___________ kcal•m^{-2}•hr^{-1}

RMR (post-meal): ___________ kcal•m^{-2}•hr^{-1}

Relative RMR change (%) = (Post-RMR – Pre-RMR)/Pre-RMR × 100% = _________%

In addition to this set of calculations from the laboratory settings, you may also use the following equation(s) to calculate an estimated RMR if you do not have access to such laboratory settings/equipment (adapted from Plowman & Smith, 1997):

Males:

RMR = 88.362 (4.799 × HT) + (13.397 × WT) – (5.677 × AGE)

Females:

RMR = 447.593 + (3.098 × HT) + (9.247 × WT) – (4.330 × AGE)

Where HT = height (centimetres), WT = weight (kilograms), AGE = age (year)

Questions (30 marks)

1. Report pre-meal RMR in $kcal \cdot m^{-2} \cdot hr^{-1}$. (5 marks)
2. Report relative RMR change (%) (pre-meal). (10 marks)
3. Based on the pre-meal RMR, what is the minimum daily expenditure of this participant? (5 marks)
4. What was the RMR in $kcal \cdot m^{-2} \cdot hr^{-1}$ after the protein meal? (5 marks)
5. Report the relative change in RMR as a result of the protein meal (%). (5 marks)

Discuss (30 marks)

1. Was the pre-meal RMR within the acceptable range? What factors could have cause the RMR to be elevated? What factors could cause the RMR to be below normal? (8 marks)
2. Would you expect the RMR to be greater in the muscular or fat individual of the same body weight? (7 marks)
3. Why does RMR decrease with age? (5 marks)
4. Define the specific dynamic action (SDA). How did the consumption of protein meal alter RMR? (10 marks)

Laboratory Session 4

Effort Sense

This laboratory work is used to determine the ratings of perceived exertion across two different submaximal treadmill run test. A modified ratings of perceived exertion (RPE) scale is used for the protocols.

Equipment needed:

- Treadmill
- Heart rate monitor and metabolic cart
- Modified RPE scale
- Recording sheet

Determination of Ratings of Perceived Exertion

In this laboratory session, there will be two trials performed by the participant on two separate days or performed in a single day with a 20-minute rest in between trials. RPE will be determined on both days using the modified OMNI RPE scale (Figure 1). This modified RPE scale will allow participants to better understand the scale and rate of perceived exertion by an individual before they commence on full validation and self-regulation laboratories in the next two chapters.

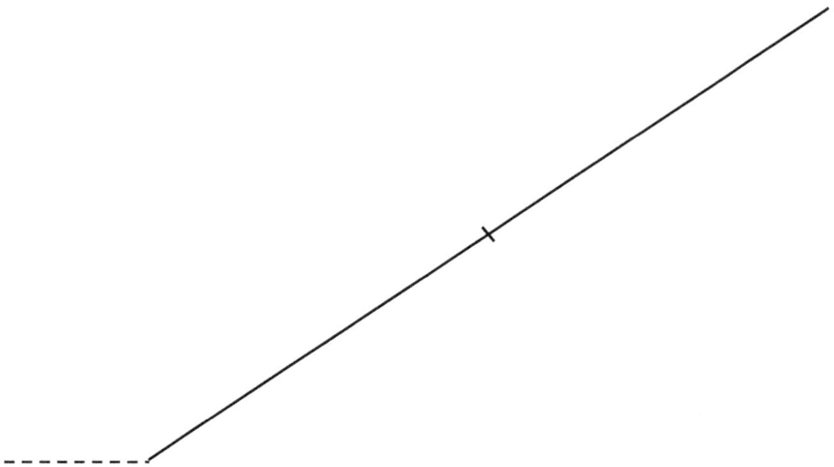

Figure 1. The modified OMNI RPE scale based on Robertson et al., 2000.

Instructions for the scale:

- How tired does your body feel during exercise?
- Instructions: We would like you to walk and then run on a treadmill for a little while. At each subsequent stage, the treadmill speed will increase. The slope in Figure 1 represents how tired you feel during the exercise. For each stage of the exercise, please use a pencil/pen to mark on the slope to tell us how tired you feel. If you are just starting out walking or jogging on the treadmill and you are not tired at all, you should make your mark at the bottom of the slope (point to the bottom of the slope on Figure 1). Now, if you are barely able to run on the treadmill because of the high speed, you will be very, very tired and you should make your mark at the top of the slope (point to the top of the slope on Figure 1). If you feel your tiredness is somewhere in between the bottom and the top of the slope, then make a mark in between these two extreme points. The midpoint of the scale is indicated with the short-dashed line in the middle (point to the midpoint on Figure 1).
- We will ask you to mark on the slope to tell us how your whole body feels. Remember, there are no right or wrong answers.

Instructions for researcher using the scale:
The researcher has to indicate where the participant points on the scale when he/she is exercising on the treadmill. Remember to indicate with a pencil/pen and also remember the stage of the treadmill test at the point of indication.

The total length of the scale from the bottom till the top of the slope will be given and thus, the researcher has to record RPE results in mm and then convert them to percentage. This will enable the data to be used for statistical purposes.

For example:
The length of the line from the bottom to the top of the slope is 100 mm. The participant's first RPE at stage 1 is marked at 15 mm. This will be indicated as the RPE occurred at the 15% point of the 100% scale.

Laboratory Session 4.1: Submaximal Till Exhaustion Protocol (Discontinuous Protocol) (Gupta & Balasekaran, 2013; Ali, Balasekaran, Hoon, & Gerald, 2017; Balasekaran et al., 2020)

Equipment needed:

1. Treadmill: set at a gradient of 1% to reflect the energy cost of running outdoors
2. Metabolic cart (to measure oxygen consumption; please ensure it is calibrated)
3. Polar heart rate monitor transmitter
4. Spirometry head gear
5. Nose clip
6. Modified OMNI RPE scale (Figure 1)
7. Safety harness
8. Water for participant

The participant will perform a discontinuous submaximal till exhaustion protocol in order to determine RPE.

Prior to the test, the participant will commence warm-up. He/she will jog an easy or slow pace on the treadmill for 3–5 minutes, followed by simple major muscle group stretches. In this session, your participant will perform a series of discontinuous treadmill runs lasting 4 minutes for each stage. The treadmill speed will increase at 0.5 $km{\cdot}h^{-1}$ with each stage, which should range between 6–14 $km{\cdot}h^{-1}$ depending on the ability of your participant and the start or end speed could be higher. Between each stage, your participant will have 4 minutes recovery. Steady state VO_2 for each stage will stabilise around the final 2 minutes of each stage. In the last minute of each stage, the participant will mark a dash on the modified OMNI RPE scale (Figure 1) to record his/her tiredness. Heart rate (HR) values will also be recorded. Exercise will continue till volitional exhaustion of the participant (HR ≥ 95% of maximal HR (max HR = 220 – age)) (For VO_{2max} criteria, refer to American College of Sports Medicine Guidelines, 2018; *Applied Physiology of Exercise* Chapter 5, Balasekaran, Govindaswamy, Lim, Boey, & Ng, 2021).

Do ensure to give at least 20 minutes rest time between the two trials (Laboratory Sessions 4.1 & 4.2) for your participant to recover if you conduct both trials on the same day.

Results can be tabulated as follows (Table 1).

Fill in the following table with your participant's submaximal till exhaustion data (remember to average the data from the final 2 minutes of each stage). (Discontinuous Protocol) (Gupta & Balasekaran, 2013; Ali, Balasekaran, Hoon, & Gerald, 2017)

Name of Participant: ______________________

Weight: ________ Height: __________ BMI: __________ Age: ________

Resting HR: _________ Predicted Max HR: __________

95% of Maximal HR: ___________

Table 1. Participant's submaximal till exhaustion data.

Stages	Treadmill Velocity ($km{\cdot}h^{-1}$)	RPE Overall (mm/%)	VO_2 ($mL{\cdot}kg^{-1}{\cdot}min^{-1}$)	VO_2 ($L{\cdot}min^{-1}$)	Heart Rate ($beats{\cdot}min^{-1}$)
1					
2					
3					
4					
5					
6					
7					
8					
9					
10					
11					
12					
13					
14					
15					
16					
17					
18					
19					
20					
21					
22					
23					
24					
25					
26					
27					
28					
29					
30					

Maximum HR: ______________________

Final stage on treadmill: __________________________

Total exercise time on treadmill: ________________________

Laboratory Session 4.2: Submaximal Till Exhaustion Protocol (Continuous Protocol) (Gupta & Balasekaran, 2013; Ali, Balasekaran, Hoon, & Gerald, 2017; Balasekaran et al., 2020)

Equipment needed:

1. Treadmill: set at a gradient of 1% to reflect the energy cost of running outdoors
2. Metabolic cart (to measure oxygen consumption; please ensure that it is calibrated)
3. Polar heart rate monitor transmitter
4. Spirometry head gear
5. Nose clip
6. Modified OMNI RPE scale (Figure 1)
7. Safety harness
8. Water for participant

The participant will perform a continuous submaximal till exhaustion protocol in order to determine RPE.

Prior to the test, the participant will commence warm-up. He/she will jog an easy or slow pace on the treadmill for 3–5 minutes, followed by simple major muscle group stretches. In this session, your participant will perform a series of continuous treadmill runs lasting 4 minutes for each stage. The treadmill speed will increase at 0.5 $km{\cdot}h^{-1}$ with each stage, which should range between 6–14 $km{\cdot}h^{-1}$ depending on the ability of your participant and the start or end speed could be higher. Steady state VO_2

for each stage will stabilise around the final 2 minutes of each stage. In the last minute of each stage, the participant will mark a dash on the modified OMNI RPE scale (Figure 1) to record his/her tiredness. Heart rate (HR) values will also be recorded. Exercise will continue till volitional exhaustion of the participant (HR ≥ 95% of maximal HR (max HR = 220 – age)) (For VO_{2max} criteria, refer to American College of Sports Medicine Guidelines, 2018; *Applied Physiology of Exercise* Chapter 5, Balasekaran, Govindaswamy, Lim, Boey, & Ng, 2021).

Results can be tabulated as follows (Table 2).

Fill in the following table with your participant's submaximal till exhaustion data (remember to average the data from the final 2 minutes of each stage). (Continuous Protocol) (Gupta & Balasekaran, 2013; Ali, Balasekaran, Hoon, & Gerald, 2017)

Name of Participant: ______________________________

Weight: ________ Height: __________ BMI: __________ Age: ________

Resting HR: _________ Predicted Max HR: __________

95% of Maximal HR: ___________

Table 2. Participant's submaximal till exhaustion data.

Stages	Treadmill Velocity ($km \cdot h^{-1}$)	RPE Overall (mm/%)	VO_2 ($mL \cdot kg^{-1} \cdot min^{-1}$)	VO_2 ($L \cdot min^{-1}$)	Heart Rate ($beats \cdot min^{-1}$)
1					
2					
3					
4					
5					
6					
7					
8					
9					
10					
11					
12					
13					
14					
15					
16					
17					
18					
19					
20					
21					
22					
23					
24					
25					
26					
27					
28					
29					
30					

Maximum HR: ____________________

Final stage on treadmill: ________________________

Total exercise time on treadmill: ______________________

Laboratory Activity (70 marks)

1. Tabulate the participant's results into an Excel spreadsheet.
2. Transfer the data to SPSS or any other statistical software and use it to perform simple statistics to determine the results.
3. Describe the pattern of rate of perceived exertion responses for both trials. (5 marks)
4. Does the modified scale with dashes on an incline without verbal descriptors or pictorials elicit a corresponding increase in RPE for workload increase for both trials? Explain your answer with reasons. (6 marks)
5. Are the dashes on the incline slope for both Laboratory Sessions 4.1 and 4.2 proportional to the increase in workload similar to the responses validated by the real OMNI RPE scale as observed in research? Explain the psycho-physio rationale for the participant to be able to indicate his/her RPE without any verbal or pictorial cues to guide them. (8 marks)
6. Are there any differences in the RPE responses when comparing between continuous and discontinuous treadmill protocols? Explain. (8 marks)
7. Use Excel to create the RPE versus VO_2 and RPE versus HR response diagrams separately for both trials. (8 marks)
8. What is the correlation between RPE versus VO_2 and RPE versus HR separately for both trials? (5 marks)
9. What is the validation criterion based on the basic principles of Borg's three effort continua model? (6 marks)
10. Did the results of the correlation conform to Borg's three effort criteria of validity? (6 marks)
11. What are the reasons for the development of the OMNI RPE scale as opposed to using Borg's RPE scale for children? (10 marks)
12. What is the validity of the OMNI RPE scales as effective tools for prescribing/monitoring appropriate exercise intensity? (8 marks)

Laboratory Session 5

Exercise Validation Trial Using the OMNI RPE Scale

The rate of perceived exertion (RPE) scale is used to determine the perceived effort by the user during any exercise. The OMNI RPE scale was developed by Robertson et al. (2000) as RPE is an excellent choice of rating scale to determine the intensity of exercise that an individual undergoes during the exercise. The OMNI RPE scale is developed based on three factors—physiological, perceptual and performance. The OMNI RPE scale is constructed on the expectation that "as the intensity of exercise performance increases, corresponding and interdependent changes occur in both the perceptual and physiological process" (Robertson et al., 2000, 2001). The interrelationship between the three factors indicates that a response in one domain provides a similar amount of information of the performance level in the other domain (Robertson et al., 2000, 2001). As such, the exercise intensity and performance level can be determined by the dynamism between the perceptual and physiological continua. Tables 1 and 2 present a list of mediators under the physiological and psychological domain.

The OMNI RPE scale has been used by coaches, physical education teachers, exercise practitioners, and researchers for field-based performance as well as laboratory works to determine the participant's work intensity (Robertson et al., 2000, 2001; Balasekaran, Loh, Govindaswamy, & Robertson, 2012; Balasekaran, Loh, Govindaswamy, & Cai, 2014; Balasekaran, Thor, Ng, & Govindaswamy, 2014; Balasekaran, Ismail, & Thor, 2015; Balasekaran, Boey, & Ng, 2018; Chia & Balasekaran, 2018; Balasekaran, Govindaswamy, Ng, & Boey, 2019; Balasekaran, Lim, Govindaswamy, Ee, & Ng, 2019; Balasekaran et al., 2019; Balasekaran, Boey, & Ng, 2020). The highly-verified scale was used across age groups, ethnicities, and genders, which

Table 1. Physiological mediators.

Respiratory — Metabolic	Peripheral	Non-Specific
Pulmonary ventilation	Metabolic acidosis (pH, lactic acid)	Hormonal regulation (catecholamines, B-endorphins)
Oxygen uptake	Blood glucose	Temperature regulation (core and skin)
Carbon dioxide production	Blood flow to muscle	Pain
Heart rate	Muscle fibre type	Cortisol and serotonin
Blood pressure	Free fatty acids	Cerebral blood flow and oxygen
	Muscle glycogen	

Table 2. Psychosocial mediators.

Classification	Factor
Emotion or mood	Anxiety Depression Extroversion Neuroticism
Cognitive function	Dissociation Self-efficacy Type A personality
Perceptual process	Pain tolerance Sensory augmentation or reduction Somatic perception
Social or situational	Music Gender of counsellor Social setting

further affirms the credibility in the field of human performance research (Robertson et al., 2000, 2001; Cai & Balasekaran, 2004; Loh & Balasekaran, 2004; Balasekaran, Loh, Govindaswamy, & Robertson, 2012; Thor & Balasekaran, 2012).

Laboratory Sessions 5 and 6 have been curated to help you better understand how and when to incorporate the use of OMNI RPE scale in performance-based research (Balasekaran, Lim, Govindaswamy, Ee, & Ng, 2019; Balasekaran et al., 2019). Prior to using the OMNI RPE scale in your session, it is important for your participant to understand how to interpret the OMNI RPE scale and how the researcher should question the participant. In order to elicit a definite answer, an orientation anchoring trial is required. This chapter aims to teach, in detail, the use of the OMNI RPE scale during laboratory testing (Figure 1).

The procedure of an orientation trial will include introduction and familiarisation of the OMNI RPE scale:

1) Use the standardised instructions below to guide the participant to rate his/her OMNI RPE (Table 3)
2) Anchoring of OMNI RPE scale (Table 4)

OMNI RPE Exercise Validation Trial

Procedure:

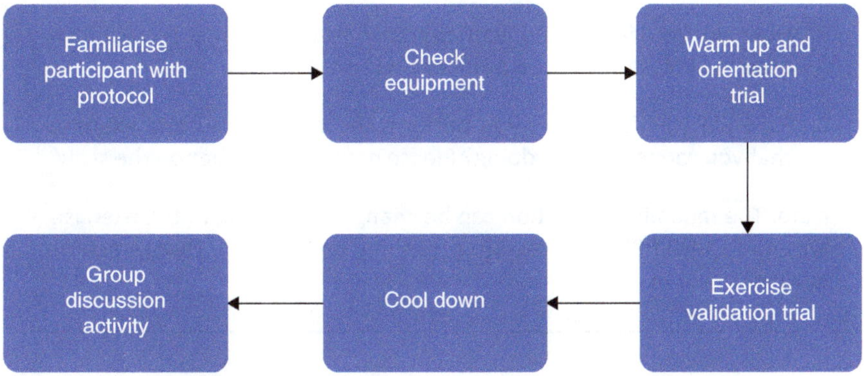

Figure 1. Procedure for OMNI rate of perceived exertion (RPE) exercise validation trial.

Table 3. OMNI rate of perceived exertion (RPE) instructions.

Standardised Instructions (bicycle ergometer has been used in this instructions but OMNI RPE and Borg RPE have been validated in cross-modals studies (Robertson et al., 1990; Pfeiffer, Pivarnik, Womack, Reeves, & Malina, 2002; Balasekaran et al., 2003))
I would like you to ride on a bicycle ergometer. Please use the numbers on this scale to tell me how your body feels when you are cycling. Please look at the person at the bottom of the hill who is just starting to ride a bicycle (*point to the left-hand picture*). If you feel like how this person looks when you are riding, the exertion will be *not tired at all*. You should point to 0. Now look at the person who is barely able to ride a bicycle to the top of the hill (*point to the right-hand picture*). If you feel like how this person looks when you are riding, you will be *very, very tired*. You should point to the number 10. If you feel like you are somewhere between *not tired at all* (0) and *very, very tired* (10), then point to a number between 0 and 10. I will ask you to point to the number that represents how your whole body feels. There are no right or wrong numbers. Use both the pictures and words to help you select the numbers. Use any of the numbers to tell us how you feel when you are riding the bicycle. Administer the following questions to the participant to ensure that the participant can comfortably and competently use the OMNI RPE scale: 1. How do you feel right now? Please point to a number on the scale. 2. How do you feel when you perform your favourite recreational activity? Please point to a number on the scale. 3. How did you feel when you perform the most exhausting exercise that you can remember doing? Please point to a number on the scale. Note: The modality and action can be changed according to the exercise type and OMNI RPE scale used (e.g. running on treadmill, change action to run).

Table 4. Orientation — anchoring of OMNI rate of perceived exertion (RPE) scale (adapted from Robertson, 2004).

Memory Anchoring of Scale
1) Ask the participant to think of a time when he/she reached a level of exertion that is equal to that depicted by the pictures at the bottom (the low-anchor point) and top (high-anchor point) of the hill in the OMNI RPE scale illustrations 2) During the exercise session, ask the participant to estimate his/her RPE by using his/her memory of the levels of exertion equal to the low and high anchors on the OMNI RPE scale

Laboratory Session 5.1: Submaximal till Exhaustion Exercise Validation Protocol (Discontinuous Protocol) (Gupta & Balasekaran, 2013; Ali, Balasekaran, Hoon, & Gerald, 2017; Balasekaran et al., 2020)

Equipment needed:

1. Treadmill: set at a gradient of 1% to reflect the energy cost of running outdoors
2. Metabolic cart (to measure oxygen consumption; please ensure that it is calibrated)
3. Polar heart rate monitor transmitter
4. Spirometry head gear
5. Nose clip
6. OMNI RPE scale (Robertson, 2004)
7. Safety harness
8. Water for participant

Prior to the test, the participant will commence warm-up. He/she will jog an easy or slow pace on the treadmill for 3–5 minutes, followed by simple major muscle group stretches. In this session, your participant will perform a series of discontinuous treadmill runs lasting 4 minutes for each stage. The treadmill speed will increase at 0.5 km•h^{-1} with each stage, which should range between 6–14 km•h^{-1} depending on the ability of your participant and the start or end speed could be higher. Between each stage, your participant will have 4 minutes recovery. Steady state VO_2 for each stage will stabilise around the final 2 minutes of each stage. In the last minute of each stage, the participant will be asked to rate his/her tiredness with reference to the OMNI RPE scale. HR values will also be recorded. Exercise will continue till volitional exhaustion of the participant (HR ≥ 95% of maximal HR (max HR = 220 – age)) (For VO_{2max} criteria, refer to American College of Sports Medicine Guidelines, 2018; *Applied Physiology of Exercise* Chapter 5, Balasekaran, Govindaswamy, Lim, Boey, & Ng, 2021).

Please note: OMNI RPE validation at ventilatory breakpoint (V_{pt}) or ventilatory threshold (VT) and lactate threshold (LT) (see Laboratory Session 10 for calculations of VT and LT) can be used as a laboratory session. In this session, OMNI RPE at V_{pt} and LT can be determined (refer to Table 6 for V_{pt} data input for V_{pt} and LT determination).

Additionally, OMNI RPE can be validated in another laboratory session using the Children's OMNI RPE scale and instructions for adolescents and children (Robertson, 2004). The treadmill protocols for adolescents (13–18 years) and children (below 13 years old) are different (age guidelines according to ACSM Guidelines (2010)).

For the adolescent's treadmill protocol, the participant can start at a walking speed of 5 km•h^{-1} and 0% gradient on the treadmill for 2 minutes before the speed starts to increase at 0.6 km•h^{-1} every 2 minutes until volitional exhaustion. In the last 30–60 seconds of each minute of exercise, the participant's Children's OMNI RPE will be recorded (Thor & Balasekaran, 2012).

For the children's treadmill protocol, the participant can start with a walking/running speed of 2.7 $km{\bullet}h^{-1}$ with 10% gradient before the speed gradually increases to 4 $km{\bullet}h^{-1}$ with 12% gradient and 5.4 $km{\bullet}h^{-1}$ with 14% gradient at 3 minutes interval. The treadmill speed will continue to increase to 6.7 $km{\bullet}h^{-1}$ with 16% gradient and 8$km{\bullet}h^{-1}$ with 18% gradient for the next 3 minutes interval, followed by 8.8 $km{\bullet}h^{-1}$ with 20% gradient. The participant will continue running until volitional exhaustion or until the last stage (Bruce, Kusumi, & Hosmer, 1973; Balasekaran, 1999). Therefore, V_{pt} can be determined for adolescents and children using these two protocols, respectively (refer to Table 6 for V_{pt} data input and Appendix A for V_{pt} calculations to determine V_{pt}).

Record the HR and OMNI RPE responses during the last minute of each treadmill stage on the recording sheet below (Table 5).

Fill in the following table with your participant's submaximal till exhaustion data (remember to average the data from the final 2 minutes of each stage).

Name of Participant: ______________________________

Weight: ________ Height: __________ BMI: __________ Age: ________

Resting HR: _________ Predicted Max HR: __________

95% of Maximal HR: ___________

Table 5. Participant's submaximal till exhaustion data.

Stages	Treadmill Velocity (km·h^{-1})	RPE Overall	VO_2 (mL·kg^{-1}·min^{-1})	VO_2 (L·min^{-1})	Heart Rate (beats·min^{-1})
1					
2					
3					
4					
5					
6					
7					
8					
9					
10					
11					
12					
13					
14					
15					

Maximum HR: ______________________

Final stage on treadmill: ________________________

Total exercise time on treadmill: ______________________

Table 6. Data on submaximal exercise (to determine V_{pt} or LT).

Stages	Treadmill Velocity ($km{\cdot}h^{-1}$)	RPE Overall	VO_2 ($mL{\cdot}kg^{-1}{\cdot}min^{-1}$)	VO_2 ($L{\cdot}min^{-1}$)	Heart Rate ($beats{\cdot}min^{-1}$)	Lactate ($mmol{\cdot}L^{-1}$)
1						
2						
3						
4						
5						
6						
7						
8						
9						
10						
11						
12						
13						
14						
15						

Laboratory Activity (40 marks)

Discuss the following questions:

1. What are the advantages and disadvantages of using RPE to monitor exercise intensity? (6 marks)
2. What influences your RPE responses during exercise? (8 marks)
3. Draw a graph and show the relationship between HR, VO_2 ($L{\cdot}min^{-1}$), and RPE. Discuss the response. (14 marks)
4. What is the validation procedure for RPE and what is the criteria for validation? (12 marks)

Optional (10 marks)

5. What is the OMNI RPE at V_{pt} or/and LT? Are they similar to Robertson's (2004) Adult OMNI RPE 6 or/and Children OMNI RPE 4 to 6 (Robertson et al., 2002; Thor & Balasekaran, 2012; Balasekaran et al., 2019)? Explain.

Laboratory Session 6

Self-Regulation of Exercise Intensity Using the OMNI RPE Scale

Self-Regulation Run Protocol (Continuous Protocol (Thor & Balasekaran, 2012))

Equipment needed:

1. Treadmill: set at a gradient of 1% to reflect the energy cost of running outdoors
2. Metabolic cart (to measure oxygen consumption; please ensure that it is calibrated)
3. Polar heart rate monitor transmitter
4. Spirometry head gear
5. Nose clip
6. OMNI RPE scale (Robertson, 2004)
7. Safety harness
8. Water for participant

The same participant who performed Laboratory Session 5 will also perform Laboratory Session 6. He/she will perform a continuous run on the treadmill to determine if he/she is able to self-regulate his/her exercise intensity and reproduce the RPE accordingly. This laboratory session can be performed on another day or on the same day after giving the participant a 20-minute rest following the first laboratory session.

Prior to the test, the participant will commence warm-up. He/she will jog an easy or slow pace on the treadmill for 3–5 minutes, followed by simple major muscle group stretches. In this session, your participant will perform a continuous 20-minute run. For the first 10 minutes, your participant will run at RPE 2. He/she will decide on a particular speed by increasing or decreasing the speed of the treadmill for the first 3 minutes. In the last 15–20 seconds of each minute, the participant will point to a number on the OMNI RPE scale (Robertson, 2004) and the researcher will record his/her tiredness. Heart rate (HR) and VO_2 ($L{\cdot}min^{-1}$) values will also be recorded. During the first 3 minutes of the run, the participant adjusts the speed to his/her overall effort level at RPE 2. After the 3 minutes, there will be no more adjustment and the participant will have to run at the self-selected speed for the next 7 minutes at RPE 2. In the last 15 seconds of each minute, during the entire 10-minute run, the participant will point to a number on the OMNI RPE scale (Robertson, 2004) and the researcher will record his/her tiredness (the RPE value should remain the same throughout. This is to give the participant and researcher assurance that he/she is exercising at the right intensity). HR and VO_2 ($L{\cdot}min^{-1}$) values will also be recorded.

After the self-regulation at RPE 2, the participant may be allowed to rest for 3 minutes on (straddle across the treadmill belt or walk at slower speed) or off the treadmill. Thereafter, the participant has to run at his/her self-selected speed at RPE 6 (can also be RPE 7) for 10 minutes. The first 3 minutes of the run at RPE 6 is where the participant adjusts the speed to his/her overall effort level at RPE 6. After the 3 minutes, there will be no more adjustments and the participant will have to run the next 7 minutes at RPE 6 at the self-selected speed. In the last 15 seconds of each minute, during the entire 10-minute run, the participant will point to a number on the OMNI RPE scale (Robertson, 2004) and the researcher will record his/her tiredness. HR and VO_2 ($L{\cdot}min^{-1}$) values will also be recorded.

In addition, to determine lactate threshold or ventilatory breakpoint (V_{pt}), the RPE at V_{pt} can be determined in Laboratory Session 5. The participant can also self-regulate RPE at V_{pt} which is usually between RPE 4 and 6 (Thor & Balasekaran, 2012; Chia & Balasekaran, 2018) and RPE 5 and 7 for adults (Balasekaran et al., 2019). This intensity can then be used

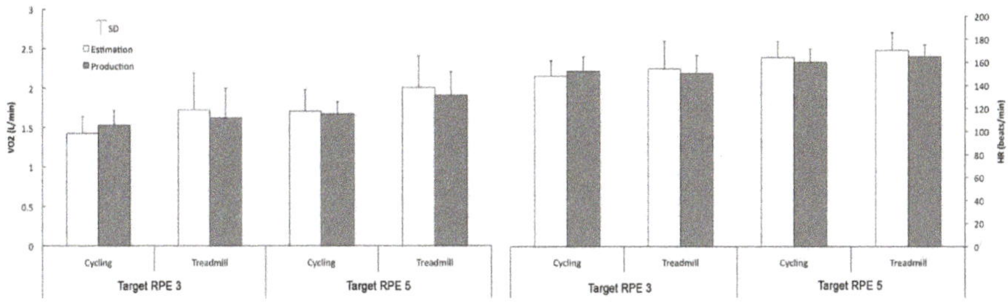

Figure 1. Prescription congruence: comparison of oxygen uptake (VO_2) and heart rate (HR) at RPE 3 and 5 with exercise validation trial in Laboratory Session 5 (note: your laboratory session data is based on RPE 2 and 6). Data are means ± SD and VO_2 data are in absolute values ($L \cdot min^{-1}$) (note: Data can be compared in $mL \cdot kg^{-1} \cdot min^{-1}$) (Thor & Balasekaran, 2012).

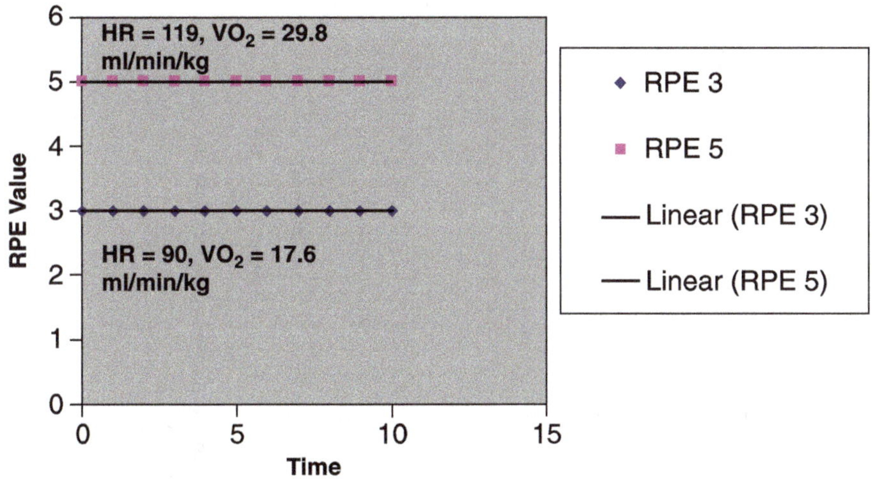

Figure 2. The above self-regulation achieved intensity discrimination for RPE 3 and 5 using hypothetical data as the HR and VO_2 data are of average values and the intensity is constant (note: your laboratory session is RPE 2 and 6. VO_2 can also be in $L \cdot min^{-1}$) (Thor & Balasekaran, 2012). Heart rate (HR, $beats \cdot min^{-1}$), oxygen uptake (VO_2, $mL \cdot kg^{-1} \cdot min^{-1}$).

during field settings. For more information on the usage of OMNI RPE for adults and children in field settings, please refer to Curricular Guide to Health-Fitness Applications in Physical Education Using the OMNI Perceived Exertion Scale developed by the authors (Balasekaran et al., 2019).

Figure 2 shows an example of exercise self-regulation by a participant for RPE 3 and 5 for 10 minutes to determine intensity discrimination (please

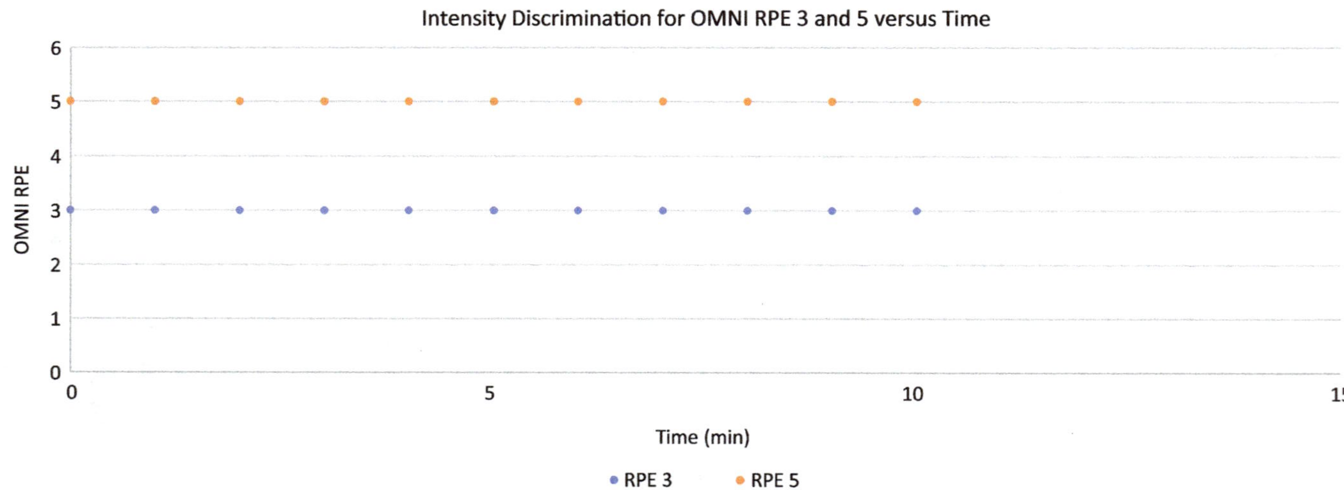

Figure 3. The above self-regulation achieved intensity discrimination for OMNI RPE 3 and 5 versus time (note: your laboratory session is based on OMNI RPE 2 and 6) (Thor & Balasekaran, 2012).

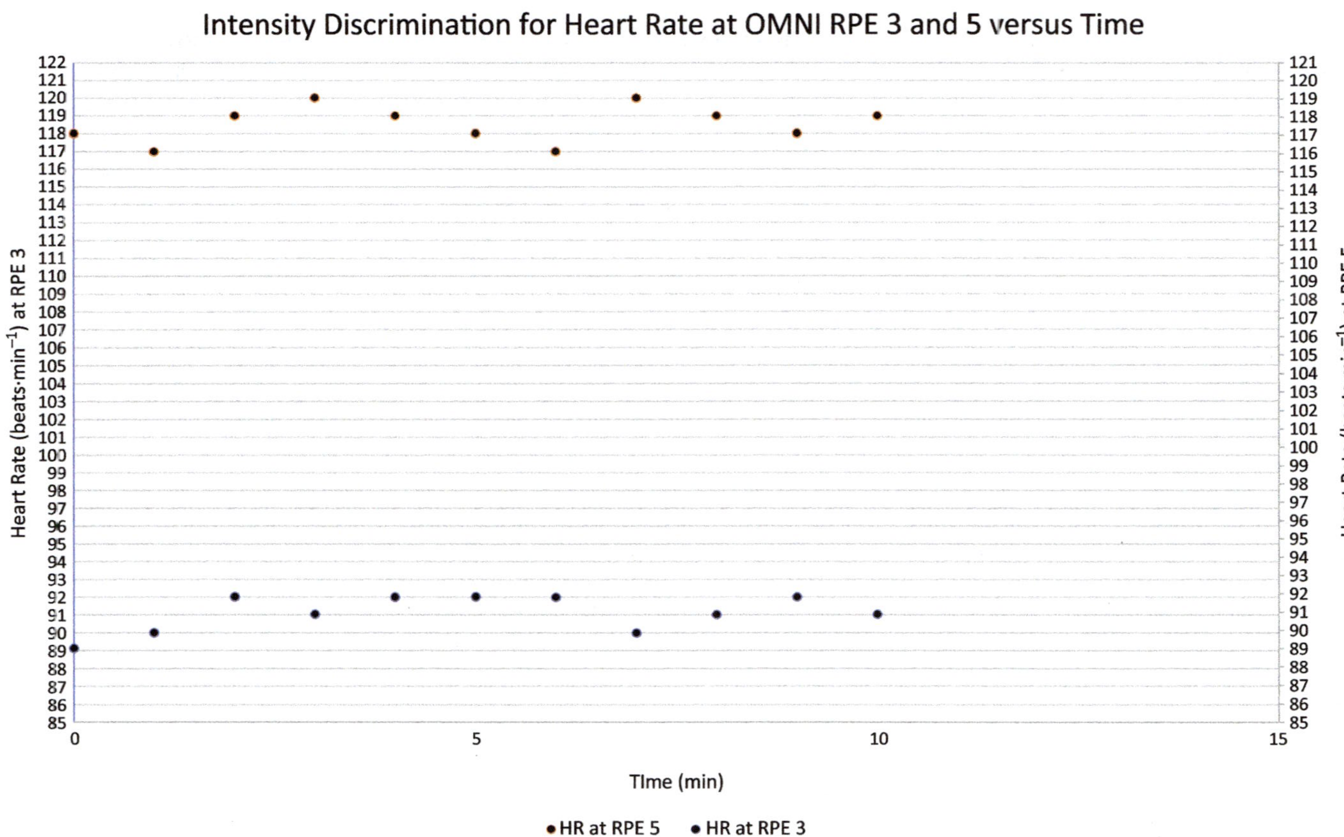

Figure 4. The above self-regulation achieved intensity discrimination for heart rate at OMNI RPE 3 (90 beats•min^{-1}) and 5 (119 beats•min^{-1}) versus time (note: your laboratory session is based on OMNI RPE 2 and 6) (Thor & Balasekaran, 2012).

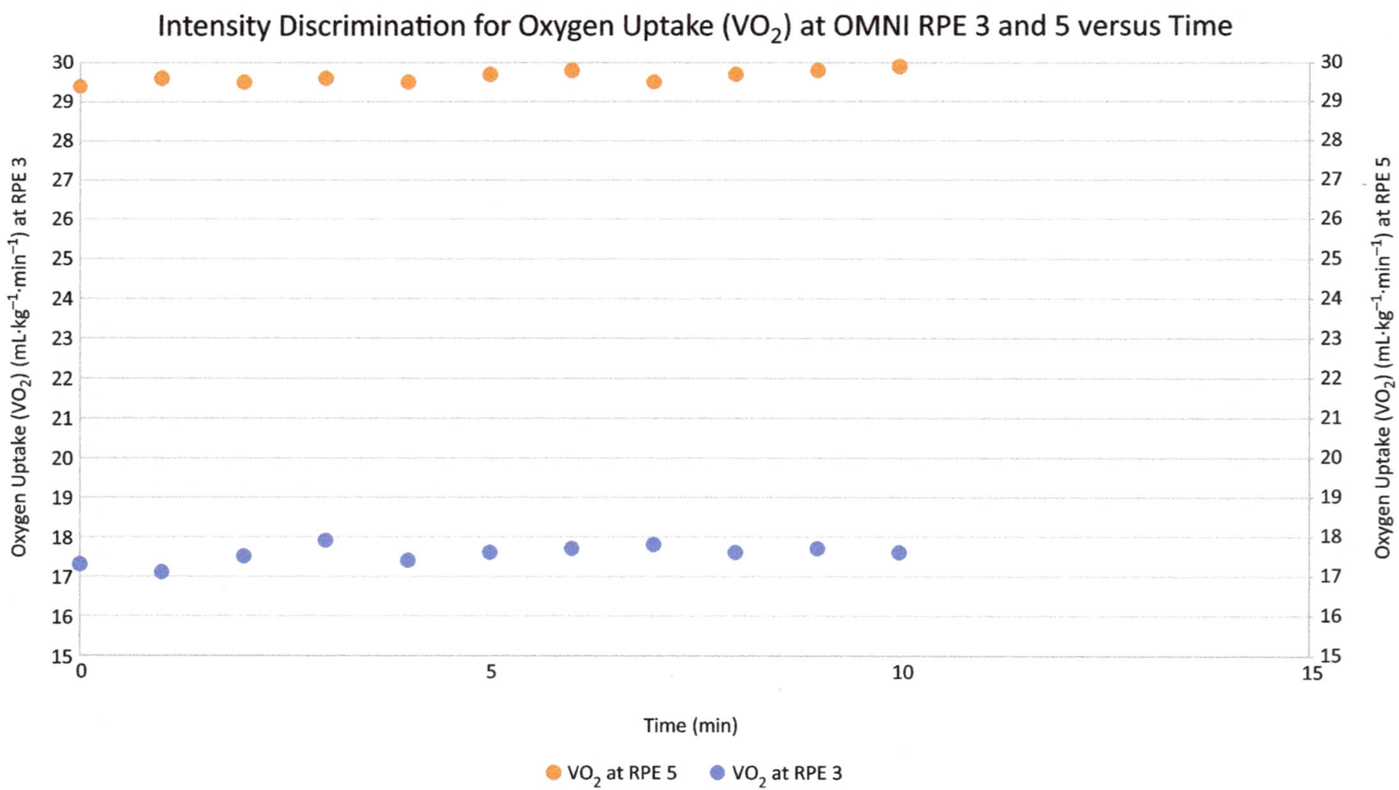

Figure 5. The above self-regulation achieved intensity discrimination for oxygen uptake (VO_2) at OMNI RPE 3 (17.6 mL•kg^{-1}•min^{-1}) and 5 (29.8 mL•kg^{-1}•min^{-1}) versus time (note: your laboratory session is based on OMNI RPE 2 and 6) (Thor & Balasekaran, 2012).

note that your laboratory session is based on RPE 2 and 6). The HR and VO_2 ($L{\cdot}min^{-1}$ or $mL{\cdot}kg^{-1}{\cdot}min^{-1}$) data of RPE 3 and 5 has to correspond to the estimation and validation trial of Laboratory Session 5 (Figure 2) (please note that your laboratory session is based on RPE 2 and 6).

Prescription Congruence for VO_2 and HR at RPE 3 and 5 (Figure 1):

- Neither VO_2 nor HR differed significantly ($p > 0.05$) between the estimation and production trials when compared at RPE 3 and 5, based on the ANOVA (trial × RPE) results. However, for laboratory purposes, the number of participants may be as low as 1 or 2, and the mean ± standard deviation is sufficient to see the differences as ANOVA is applicable for a large number of participants. This indicates prescription congruence by the participant.

Intensity Discrimination for RPE 3 and 5 (Figures 2, 3, & 4):

- Oxygen uptake and HR were significantly higher ($p < 0.01$) between self-regulated production trials (RPE 5 versus RPE 3) by 12.2 $mL{\cdot}kg^{-1}{\cdot}min^{-1}$ (29.8 $mL{\cdot}kg^{-1}{\cdot}min^{-1}$–17.6 $mL{\cdot}kg^{-1}{\cdot}min^{-1}$) and 29 $beats{\cdot}min^{-1}$ (119 $beats{\cdot}min^{-1}$–90 $beats{\cdot}min^{-1}$). This indicates intensity discrimination by the participant.

Results can be tabulated as follows (Tables 1 & 2).

Fill in the following table with your participant's self-regulation run test data at RPE 2.

Name of Participant: ______________________________

Weight: ________ Height: __________ BMI: __________ Age: ________

Resting HR: _________

Table 1. Participant's self-regulation run protocol (continuous protocol) data.

Stages/ min	Treadmill Velocity ($km \cdot h^{-1}$)	RPE Overall	VO_2 ($mL \cdot kg^{-1} \cdot min^{-1}$)	VO_2 ($L \cdot min^{-1}$)	Heart Rate ($beats \cdot min^{-1}$)
1					
2					
3					
4					
5					
6					
7					
8					
9					
10					

Fill in the following table with your participant's self-regulation run test data at RPE 6.

Name of Participant: ________________________________

Weight: ________ Height: __________ BMI: __________ Age: ________

Resting HR: _________

Table 2. Participant's self-regulation run protocol (continuous protocol) data.

Stages/ min	Treadmill Velocity ($km{\cdot}h^{-1}$)	RPE Overall	VO_2 ($mL{\cdot}kg^{-1}{\cdot}min^{-1}$)	VO_2 ($L{\cdot}min^{-1}$)	Heart Rate ($beats{\cdot}min^{-1}$)
1					
2					
3					
4					
5					
6					
7					
8					
9					
10					

Laboratory Activity (50 marks)

Discuss the following questions:

1. Can the participant self-regulate his/her exercise intensity? Was the participant able to sustain exercise at OMNI RPE 2 and 6? Explain with physiological and perceptual reasons. (6 marks)
2. Compare the physiological variables (HR and VO_2) at OMNI RPE 2 and 6. Do they correlate? (6 marks)
3. How can a physical education teacher, coach, sports practitioner, athlete, etc. use the OMNI RPE scale during exercise or training? (8 marks)
4. How does training or exercising at ventilatory breakpoint (V_{pt}), which occurs at OMNI RPE 4 to 6 for children and OMNI RPE 5 to 7 for adults, help an individual? (10 marks)
5. Did the self-regulation of OMNI RPE at 2 and 6 achieve **prescription congruence**? Explain with the use of a bar graph similar to Figure 1 and with statistics. (10 marks)
6. Did the self-regulation of OMNI RPE at 2 and 6 achieve **intensity discrimination**? Explain with the use of a line graph similar to Figures 3, 4, and 5 with statistics. (10 marks)

Laboratory Session 7

Oxygen Kinetics: Maximally Accumulated Oxygen Deficit (MAOD) to Determine Energy System Contribution During 1,500-m Run & Excess Post-Exercise Consumption (EPOC)

The immediate source of energy for muscle contraction comes from the hydrolysis of adenosine triphosphate (ATP). As ATP exists in very low concentrations in the muscle, and regulatory mechanisms appear to prevent its complete degradation, the body has evolved well-regulated chemical pathways to regenerate ATP to allow muscle contractions to continue. There are 3 distinct yet closely integrated processes that operate together to satisfy energy requirements of the muscle (Figures 1 & 2) (Refer to *Applied Physiology of Exercise* Chapters 2, 3, & 4, Balasekaran, Govinadaswamy, Lim, Boey, & Ng, 2021).

ATP Phosphocreatine (ATP-PCr)

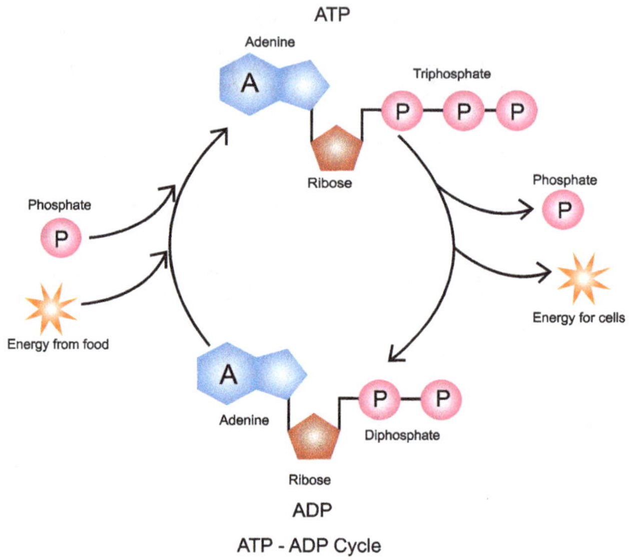

Figure 1. ATP-PCr energy process.

Aerobic and Anaerobic

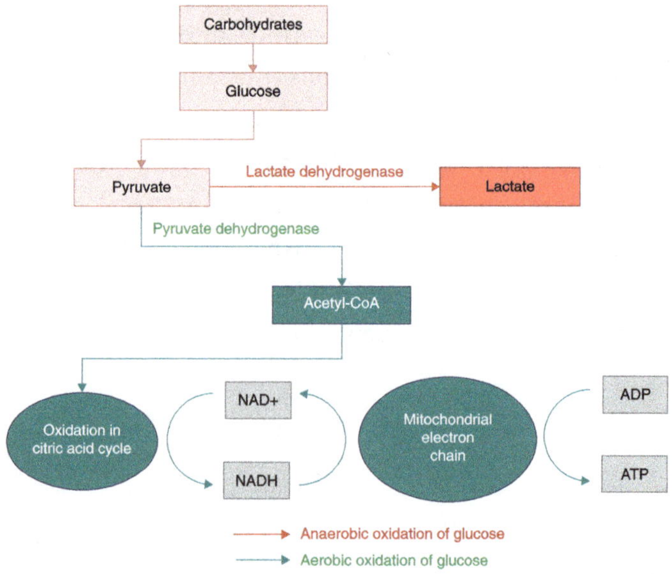

Figure 2. Aerobic and anaerobic energy processes.

Maximally Accumulated Oxygen Deficit (MAOD)

- The capacity of a person's ability to regenerate ATP from PCr, adenosine diphosphate (ADP), and glycolysis.
- Although difficult to measure, an accepted method for estimating anaerobic capacity is the **accumulated oxygen deficit** (AOD).
- AOD is larger in sprint-trained athletes than endurance-trained athletes

Recovery from Exercise: Metabolic Responses

- Excess Post-Exercise Oxygen Consumption (EPOC):
  - Elevated VO_2 for several minutes immediately following exercise
  - Was formerly known as oxygen debt
- "Fast" portion of EPOC:
  - Resynthesis of stored PCr
  - Replacing muscle and blood O_2 stores
- "Slow" portion of EPOC:
  - Elevated body temperature and catecholamines
  - Conversion of lactic acid to glucose (gluconeogenesis)

Please note that the next Laboratory Session 8 will determine "fast" portion of EPOC and "slow" portion of EPOC (Laboratory Session 8 Figures 1 & 4).

Oxygen (O_2) uptake increases exponentially at the commencement of exercise due to the time taken for the respiratory, circulatory, and metabolic processes within the muscle to adapt to the increased O_2 demand. Despite the initial increase in O_2 uptake, the actual O_2 uptake in the initial minutes of exercise is less than that required at a steady state. This is the result of the O_2 uptake at the mouth being behind the actual O_2 demand at the skeletal level until a steady state is reached (Gupta & Balasekaran, 2013). Initially, exercise scientists called this oxygen deficit, which is replaced by oxygen debt after exercise is finished. However, O_2 debt is not equal to O_2 deficit, and thus the idea of using O_2 debt as an indication of anaerobic energy has lost its direction over the years. Appropriate terms like EPOC ("fast" portion of O_2 debt and "slow" portion of O_2 debt) replaced oxygen debt as O_2 deficit is not equal to EPOC (Laboratory Session 8 Figures 1 & 4).

A time lag exists in the acceleration of actual O_2 uptake to a steady state. This results in insufficient availability of O_2 for metabolism within the working muscles. The energy required for ATP production, above which can be accounted for by measured O_2 uptake, is covered by three immediate, yet limited, sources: the breakdown of PCr and ATP in the working muscle, and the breakdown of glycogen to lactate, which are seen as the 'true' anaerobic sources of energy (Figures 1 & 2). Further energy is derived from body O_2 stores consisting of: (1) O_2 bound to haemoglobin and myoglobin; (2) O_2 dissolved in the body fluids; and (3) O_2 present in the lungs.

The difference between the steady state O_2 uptake and the actual O_2 uptake at the commencement of exercise has been defined as O_2 deficit. Traditional O_2 deficit hypotheses saw it as the result of limited O_2 transport. O_2 deficit is now seen as a consequence of metabolic processes that are necessary to improve peripheral O_2 utilisation. Therefore, the incurred O_2 deficit is a measure of the anaerobic energy release within the working muscles. This corresponds to phosphagen depletion in the working muscle, the net amount of lactate produced via glycolytic regeneration of ATP before reaching a steady state, and the small changes in the body's O_2 stores.

O_2 deficit is known to be proportional to submaximal exercise intensity. The greater O_2 deficit at higher submaximal intensities implies a greater reliance on anaerobic metabolism during the initial stages of exercise due to slower O_2 kinetics with increasing exercise intensity. Short, intense bouts of exercise, which lead to exhaustion within 2 minutes, rely mainly on the immediate (ATP-PC) and short-term (anaerobic glycolysis) energy systems (Figures 1 & 2). Unfortunately, few standardised procedures exist for the accurate estimation of anaerobic energy release, with no uniformly agreed procedure for the measurement of anaerobic capacity.

While the functional capacity of the aerobic energy system (maximal oxygen uptake (VO_{2max})) is relatively easy to measure, quantification of anaerobic energy release is difficult. Anaerobic energy release involves intracellular mechanisms, which are difficult to explicate and accurately quantify. Concerns exist over the quantification of anaerobic energy release via the O_2 debt method. Measured changes in muscle lactate and PCr concentration have also been hypothesised as quantitative measures of anaerobic capacity. These methods are rarely used because of their invasive nature, and unless the working muscle mass is known, they provide only an estimate of anaerobic energy release.

Tests such as the 30-second Wingate Test have been shown to be too long to measure ATP-PC energy system and too short to fully exhaust the anaerobic glycolysis energy systems, so its value to determine anaerobic work capacity is limited (Withers et al., 1991; Refer to *Applied Physiology of Exercise* Chapter 1, Balasekaran, Govinadaswamy, Lim, Boey, & Ng, 2021). For the purpose of this laboratory, we will determine the relative contributions of the anaerobic and aerobic energy systems to a 1,500-m run. To do this, we will use the maximally accumulated oxygen deficit (MAOD) technique (Medbø et al., 1988) which was adapted and developed by Spencer and Gastin (2001) from the University of Ballarat, Australia. They concluded that the contributions of the aerobic and anaerobic systems in the laboratory would be the same as in a track race situation. In this laboratory, we will assess this assumption. MAOD deficit has been proposed as an accurate measure of anaerobic capacity (Balasekaran et al., 2021). Valid determination of O_2 deficit as a measure of anaerobic capacity requires enough time for maximal glycolytic flux to occur, while being short enough so that the resynthesis of PCr and lactate catabolism do not take place, or are minimal before exhaustion. Developmental work has seen the difference between the actual O_2 uptake and estimated or theoretical O_2 cost, with short-duration, exhaustive exercise being termed MAOD.

The MAOD method has been proposed as a precise, yet non-invasive, measure of anaerobic energy release during cycling and running. The AOD is based on the linear relationship that exists between submaximal exercise intensities and O_2 demand (Figure 5). At supramaximal intensities, the O_2 demand is estimated by an extrapolation of this relationship with the anaerobic contribution being taken as the difference between O_2 demand and uptake (Vikneswaran & Balasekaran, 2003; Yong & Balasekaran, 2003; Balasekaran et al., 2021).

The validity of the MAOD method is based on three assumptions. Firstly, ATP utilisation begins at the onset of exercise and continues at a constant rate at any constant submaximal exercise intensity for periods up to 10 minutes in duration. Secondly, the steady state O_2 uptake, which increases linearly with exercise intensity (Figure 5), reflects ATP utilisation during the exercise. Finally, steady state energy requirements, as measured by O_2 uptake, increase linearly at both submaximal and supramaximal (above VO_{2max}) exercise intensities. Therefore, via the measurement of the accumulated O_2 uptake, the AOD for a given exercise intensity is equal to the calculated O_2 demand minus the actual O_2 uptake. Despite these assumptions, the MAOD method has been shown to be an accurate and reliable indicator of energy demands above those that can be met by aerobic processes (Graham & McLellan, 1989) (Figures 3 & 4).

The MAOD method has indicated that for exhausting exercise lasting between 30 seconds and 3 minutes, the total energy release and work done is the sum of a (aerobic) component proportional to duration and a constant

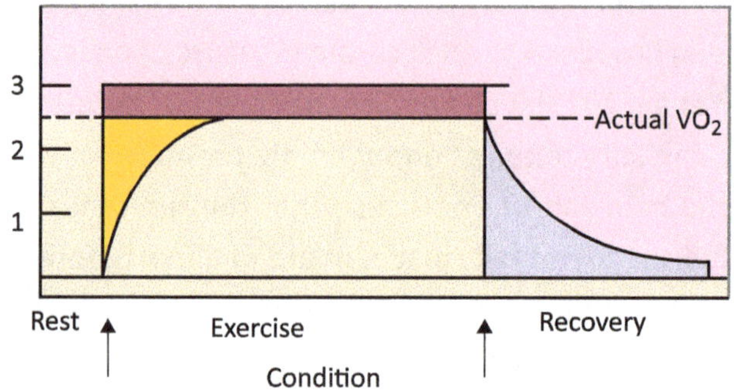

Figure 3. Maximal accumulated oxygen deficit (MAOD): Area between the curve of the O_2 demand and the curve of the actual O_2 uptake (adapted from Medbø et al., 1988).

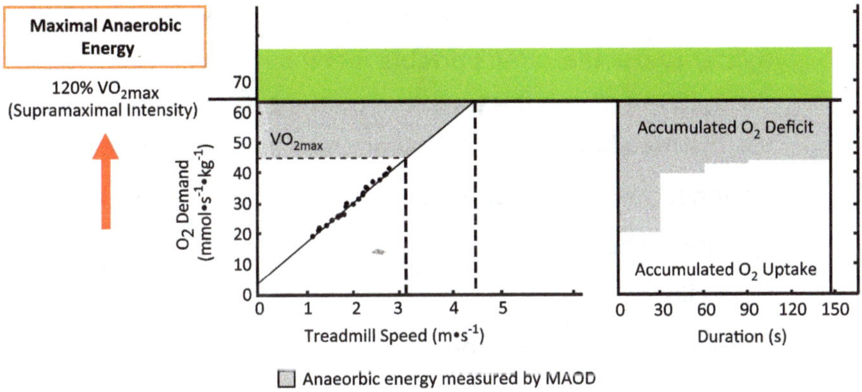

Figure 4. Based on estimated supramaximal intensity to measure maximal anaerobic energy of an individual (MAOD) (adapted from Gupta & Balasekaran, 2013).

(anaerobic) addition. The AOD has been shown to increase with the duration of supramaximal exercise. Almost three-fold increases have been observed when exercise duration increases from 15 seconds to 2 minutes, with negligible increases after that up to 16 minutes. After 60 seconds of maximal work, aerobic energy sources provide approximately 50% of the total energy requirement, and after 2 minutes, the aerobic contribution increases to ± 60% (Medbø et al., 1988; Vikneswaran & Balasekaran, 2003; Yong & Balasekaran, 2003).

Laboratory Session 7.1: Submaximal Efficiency Tests and VO_{2max} Determination (Gupta & Balasekaran, 2013; Ali, Balasekaran, Hoon, & Gerald, 2017; Balasekaran et al., 2020)

Equipment needed:

1. Treadmill: set at a gradient of 1% to reflect the energy cost of running outdoors
2. Cosmed K4b2 or $K4b^2$ (they can be used interchangeably) pre-calibrated (or other variant models of the portable metabolic

machine) or Parvomedics metabolic cart (or other metabolic carts) (better to use the same portable metabolic machine as used for the 1,500-m run on the track and on the treadmill. You may also use the metabolic cart but accuracy may be compromised.)

3. Polar heart rate monitor transmitter
4. Spirometry head gear
5. Nose clip
6. OMNI RPE scale (Robertson, 2004)
7. Safety harness
8. Water for participant

Prior to the test, the participant will commence warm-up. He/she will jog an easy or slow pace on the treadmill for 3–5 minutes, followed by simple major muscle group stretches. In this session, your participant will perform a series of discontinuous treadmill runs lasting 4 minutes for each stage. The treadmill speed will increase at 0.5 $km{\cdot}h^{-1}$ with each stage, which should range between 6–14 $km{\cdot}h^{-1}$ depending on the ability of your participant and the start or end speed could be higher. Between each stage, your participant will have 4 minutes recovery. During the recovery stage, the researcher* will take a finger-prick blood sample from the participant to measure blood lactate. Once the blood sample is collected in a small tube (e.g. microvettes etc.), measure blood lactate results immediately using the blood lactate machine (portable blood lactate analysers may not indicate accurate results).

Steady state VO_2 for each stage will stabilise around the final 2 minutes of each stage. This will help you develop a linear relationship between VO_2

and treadmill velocity (Figure 5), which will be extrapolated to estimate O_2 cost (energy demand) during a simulated 1,500-m treadmill run. In the last minute of each stage, the participant will point to a number on the OMNI RPE scale (Robertson, 2004) and the researcher will record his/her OMNI RPE values. HR values will also be recorded. The exercise will continue till there is a sharp increase in blood lactate values, following which the test will terminate.

*Researcher to wear surgical gloves during test for hygiene purposes. Dispose of laboratory consumables (e.g. finger prick needle, used tissue, etc.) appropriately into the respective disposal bins according to your laboratory's rules and regulations.

Laboratory Session 7.1 Results (Table 1)

Fill in the following table with your participant's submaximal data (remember to average the data from the final 2 minutes of each exercise stage).

Name of Participant: ______________________

Weight: ________ Height: __________ BMI: __________ Age: ________

Resting HR: _________ Predicted Max HR: __________

95% of Maximal HR: ___________

Table 1. Participant's submaximal data.

Stages	Time	Treadmill Speed (km•h^{-1})	VCO_2 (L•min^{-1})	VO_2 (L•min^{-1})	VO_2 (mL•kg^{-1}•min^{-1})	RER	VE (L•min^{-1})	Heart Rate (beats•mins^{-1})	OMNI RPE	Lactate (mmol•L^{-1})
1										
2										
3										
4										
5										
6										
7										
8										
9										
10										
11										
12										
13										
14										
15										

(*Continued*)

Table 1. ***(Continued)***

Stages	Time	Treadmill Speed (km•h^{-1})	VCO_2 (L•min^{-1})	VO_2 (L•min^{-1})	VO_2 (mL•kg^{-1}•min^{-1})	RER	VE (L•min^{-1})	Heart Rate (beats•mins^{-1})	OMNI RPE	Lactate (mmol•L^{-1})
16										
17										
18										
19										
20										
21										
22										
23										
24										
25										
26										
27										
28										
29										
30										

Using Excel, it is possible to plot the linear relationship between submaximal treadmill speed and O_2 uptake using a scatter graph. For examples, see Table 3 and Figure 5.

Note: Weight of the portable metabolic machine can be corrected for the total load (body weight plus weight of portable metabolic machine) (Reis & Miguel, 2007).

Laboratory Session 7.2 (Gupta & Balasekaran, 2013; Balasekaran et al., 2020) (Table 2)

After approximately 20 minutes, recalibrate the metabolic cart or portable metabolic machine and allow your participant to stretch his/her major muscle groups again. Once the participant is geared up with the necessary equipment for the metabolic measurement, including HR, start the VO_{2max} test on your participant. Start the treadmill speed at the same speed as the 9th submaximal treadmill run/10 $km{\cdot}h^{-1}$ (depending on your participant's start speed). Increase the treadmill velocity by 1 $km{\cdot}h^{-1}$ for 6 minutes (to 15 $km{\cdot}h^{-1}$) (depending on your participant's start speed) at every minute. If your participant is still running, increase the treadmill gradient by 2 $\%{\cdot}min^{-1}$ while keeping speed constant until volitional exhaustion of the participant (HR ≥ 95% of maximal HR (max HR = 220 – age)) (for VO_{2max} criteria, refer to American College of Sports Medicine Guidelines, 2018; *Applied Physiology of Exercise* Chapter 5, Balasekaran, Govindaswamy, Lim, Boey, & Ng, 2021). Encourage your participant to continue running for as long as he/she possibly can.

Laboratory Session 7.3: 1,500-m Run on the Track

Equipment needed:

1. Cosmed $K4b^2$ pre-calibrated (or other variant models of the portable metabolic machine)
2. Laptop to download data from Cosmed
3. Polar heart rate monitor transmitter
4. Stopwatch
5. Water for participant

Prior to the test, the participant will commence warm-up. He/she will jog an easy or slow pace on the track for 3–5 minutes, followed by simple major muscle group stretches. Next, the participant will complete

Laboratory Session 7.2 Results (Table 2)

In the table below, record the participant's data from the VO_2 maximal test.

Table 2. Participant's data from VO_{2max} test.

Stages	Time	Treadmill Speed ($km \cdot h^{-1}$)	VCO_2 ($L \cdot min^{-1}$)	VO_2 ($L \cdot min^{-1}$)	VO_2 ($mL \cdot kg^{-1} \cdot min^{-1}$)	RER	VE ($L \cdot min^{-1}$)	Heart Rate ($beats \cdot min^{-1}$)

Post-blood lactate values immediately after termination of test: _____ $mmol \cdot L^{-1}$

Note: Weight of the portable metabolic machine can be corrected for the total load (body weight plus weight of portable metabolic machine) (Reis & Miguel, 2007).

Table 3. Sample data for velocity and O_2 consumption.

Vel (km•h^{-1})	10	11	12	13	14
VO_2 (L•min^{-1})	2.291	2.559	2.799	3.044	3.451

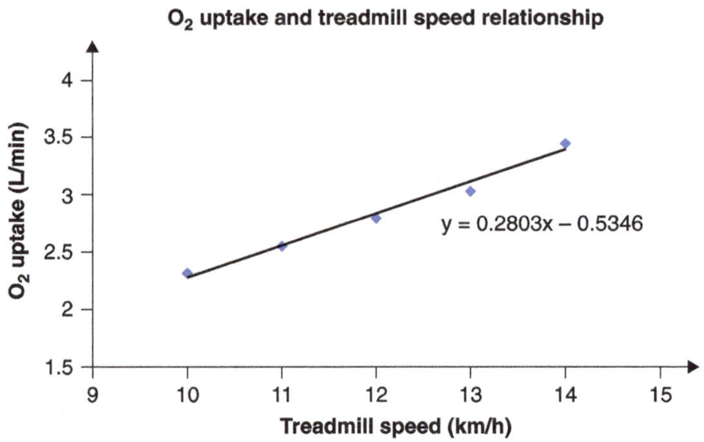

Figure 5. Example of linear regression graph between O_2 uptake (L•min^{-1}) and treadmill speed (km•h^{-1}) using hypothetical data.

Now develop your own graph and relationship between O_2 uptake and submaximal treadmill speed (Appendix A, Figure B).

Submaximal Relationship Equation = ______________________________

80 m × 4 stridings as part of his warm-up. The participant will use the Cosmed K4b^2 (or other variant models of the portable metabolic machine) and wear it during the duration of the simulated 1,500-m run. The researcher will help the participant put on the portable metabolic machine and heart rate (HR) strap. Have the transmitter connected to the Cosmed so that you can automatically download data from your participant as he/she runs. In this run, your participant has to complete the distance as fast as he/she can. In order to get the most accurate results, it is important that he/she does not jog for the first 1,100 m, then sprint the final lap. He/she has to pace well for the entire 1,500-m.

Using the relationship between O_2 uptake and the total time for the participant to complete the 1,500-m run, it is possible, with a series of equations, to determine (Figure 5):

1. The average velocity of the run
2. The calculated O_2 cost of running at that speed

Using **an example** of a 1,500-m running time of 5 minutes 55 seconds, we can calculate the average velocity and O_2 cost of the run using the following equations (Balasekaran et al., 2021):

Average Velocity: → 5 minutes 55 seconds
→ 355 seconds per 1 km
→ 1500 m is 1.5 km
→ 355/1.5 = 236.7 seconds per km
→ 4.23 km/s (reciprocal of 236.7 will give km in seconds)
→ 4.23 km/s × 3600 per hour
→ 15.23 $km{\cdot}h^{-1}$

Average velocity is 15.23 $km{\cdot}h^{-1}$

Calculated O_2 cost: $y = 0.2803x - 0.5346$ (Figure 5, hypothetical equation)
If $x = 15.23$
$y = 0.2803(15.23) - 0.5346$
Calculated O_2 cost is 3.74 $L{\cdot}min^{-1}$

Remember that O_2 deficit is the difference between estimated O_2 uptake (calculated O_2 cost) and actual O_2 uptake. At the completion of the test, average the data from the Cosmed or other portable metabolic machines into 10-second periods and then complete the following table (Table 4).

The O_2 deficit is then calculated as the difference between calculated O_2 cost and the actual VO_2 (for every 10-second period).

O_2 deficit = 3.74 $L{\cdot}min^{-1}$ (calculated O_2 cost) – Actual VO_2 ($L{\cdot}min^{-1}$) (for every 10-second period)

The O_2 deficit is then utilised in the calculation for each individual to determine his/her energy system contribution.

% aerobic contribution = (actual VO_2/calculated O_2 cost) x 100
% anaerobic contribution = (O_2 deficit/calculated O_2 cost) x 100

% aerobic contribution can also be determined by 100% – % anaerobic contribution.

(Balasekaran et al., 2021)

Laboratory Session 7.4: 1,500-m Run on the Treadmill

Equipment needed:

1. Cosmed K4b[2] pre-calibrated (or other variant models of the portable metabolic machine) or Parvomedics metabolic cart (or other metabolic carts) (better to use the same portable metabolic machine as used for the 1,500-m run on the track. You may also use the metabolic cart but accuracy may be compromised.)
2. Laptop to download data from Cosmed (if using Cosmed)
3. Polar heart rate monitor transmitter
4. Treadmill
5. Stopwatch
6. Water for participant

If possible, conduct this final test on the same treadmill as your previous two tests. Prior to the test, the participant will commence warm-up. He/she will jog an easy or slow pace on the treadmill for 3–5 minutes, followed by simple major muscle group stretches. Similar to the submaximal efficiency and VO_{2max} tests, the treadmill gradient should be set at 1%. The treadmill speed should be set at the average speed previously determined for the 1,500-m track run. This will give you a

Table 4. **1,500-m track run results.**

Time (min:sec)	Calculated O_2 Cost ($L{\cdot}min^{-1}$)	Actual VO_2 ($L{\cdot}min^{-1}$)	O_2 Deficit ($L{\cdot}min^{-1}$)	Anaerobic (%)	Aerobic (%)	Heart Rate ($beats{\cdot}min^{-1}$)
0:00–0:10						
0:11–0:20						
0:21–0:30						
0:31–0:40						
0:41–0:50						
0:51–1:00						
1:01–1:10						
1:11–1:20						
1:21–1:30						
1:31–1:40						
1:41–1:50						
1:51–2:00						
2:01–2:10						
2:11–2:20						
2:21–2:30						
2:31–2:40						
2:41–2:50						
2:51–3:00						
3:01–3:10						
3:11–3:20						
3:21–3:30						

(*Continued*)

Table 4. ***(Continued)***

Time (min:sec)	Calculated O_2 Cost ($L \cdot min^{-1}$)	Actual VO_2 ($L \cdot min^{-1}$)	O_2 Deficit ($L \cdot min^{-1}$)	Anaerobic (%)	Aerobic (%)	Heart Rate ($beats \cdot min^{-1}$)
3:31–3:40						
3:41–3:50						
3:51–4:00						
4:01–4:10						
4:11–4:20						
4:21–4:30						
4:31–4:40						
4:41–4:50						
4:51–5:00						
5:01–5:10						
5:11–5:20						
5:21–5:30						
5:31–5:40						
5:41–5:50						
5:51–6:00						
6:01–6:10						
6:11–6:20						
6:21–6:30						
6:31–6:40						
6:41–6:50						
6:51–7:00						

Note: Weight of the portable metabolic machine can be corrected for the total load (body weight plus weight of portable metabolic machine) (Reis & Miguel, 2007).

constant velocity test as used in the methodology of Spencer and Gastin (2001). Have your participant step onto the moving treadmill (i.e. 10–12 km•h^{-1} is the speed for the participant whose calculated velocity was 15.23 km•h^{-1}) and then increase treadmill speed in the shortest possible time to the required calculated velocity. Upon reaching the desired speed, simultaneously start a stopwatch and the Cosmed K4b2 (or other variant models of the portable metabolic machine) or Parvomedics metabolic cart (or other metabolic carts) using the remote start on the computer. The test is completed when the participant hit the 1,500-m track time (i.e. 5 minutes 55 seconds, see above example). Extract the run data using 10-second intervals and then complete Table 5.

Table 5. 1,500-m treadmill run results.

Time (min:sec)	Calculated O_2 Cost ($L{\cdot}min^{-1}$)	Actual VO_2 ($L{\cdot}min^{-1}$)	O_2 Deficit ($L{\cdot}min^{-1}$)	Anaerobic (%)	Aerobic (%)	Heart Rate ($beats{\cdot}min^{-1}$)
0:00–0:10						
0:11–0:20						
0:21–0:30						
0:31–0:40						
0:41–0:50						
0:51–1:00						
1:01–1:10						
1:11–1:20						
1:21–1:30						
1:31–1:40						
1:41–1:50						
1:51–2:00						
2:01–2:10						
2:11–2:20						
2:21–2:30						
2:31–2:40						
2:41–2:50						
2:51–3:00						
3:01–3:10						
3:11–3:20						

(*Continued*)

Table 5. ***(Continued)***

Time (min:sec)	Calculated O_2 Cost ($L \cdot min^{-1}$)	Actual VO_2 ($L \cdot min^{-1}$)	O_2 Deficit ($L \cdot min^{-1}$)	Anaerobic (%)	Aerobic (%)	Heart Rate ($beats \cdot min^{-1}$)
3:21–3:30						
3:31–3:40						
3:41–3:50						
3:51–4:00						
4:01–4:10						
4:11–4:20						
4:21–4:30						
4:31–4:40						
4:41–4:50						
4:51–5:00						
5:01–5:10						
5:11–5:20						
5:21–5:30						
5:31–5:40						
5:41–5:50						
5:51–6:00						
6:01–6:10						
6:11–6:20						
6:21–6:30						
6:31–6:40						
6:41–6:50						
6:51–7:00						

Note: Weight of the portable metabolic machine can be corrected for the total load (body weight plus weight of portable metabolic machine) (Reis & Miguel, 2007).

Laboratory Session 7 Data Analysis (30 marks)

1. Draw line graphs which plot the relative contribution of the aerobic and anaerobic system (%) in 10-second intervals for both the track and treadmill 1,500-m runs. (10 marks) (see Figure 6 below)

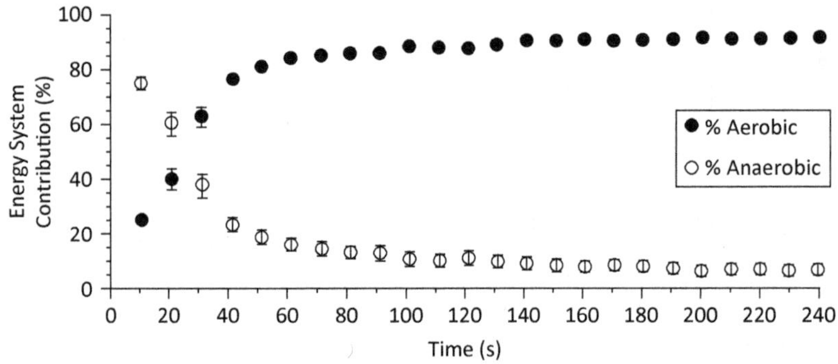

Figure 6. Relative contribution of the aerobic and anaerobic energy systems in 10-second intervals for 1,500-m run (adapted from Spencer, Gastin, & Payne, 1996).

2. Draw column charts that plot O_2 deficit and O_2 uptake in 10-second intervals for both the track and treadmill 1,500-m runs. (6 marks) (see Figure 7 below)

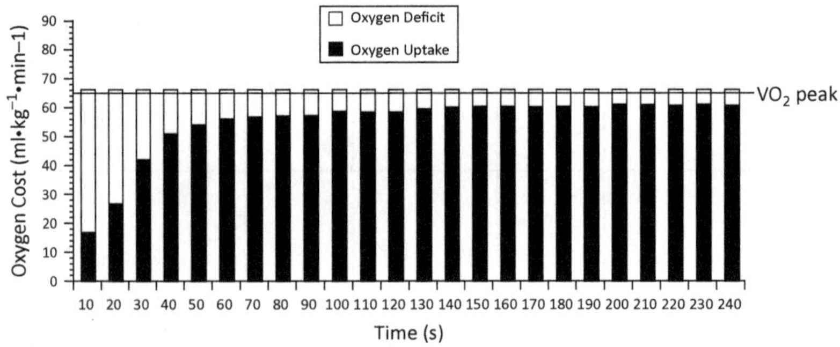

Figure 7. Oxygen deficit and oxygen uptake in 10-second intervals for 1,500 m (adapted from Spencer, Gastin, & Payne, 1996)

3. Calculate the total aerobic and anaerobic contributions to the total O_2 cost of the 1,500-m track and treadmill runs, plotting them both on the same graph. (14 marks) (see Figure 8 below) (Note: for your laboratory session, you only need to plot the 1,500 m bar graph)

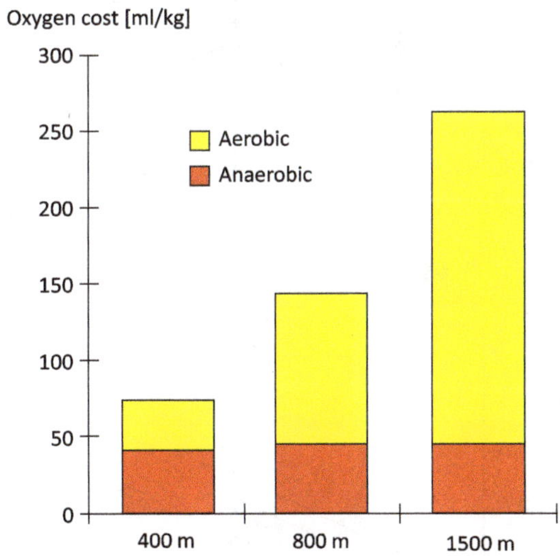

Figure 8. Aerobic and anaerobic contribution to the total oxygen cost of 400-m, 800-m, and 1,500-m runs (adapted from Spencer, Gastin, & Payne, 1996).

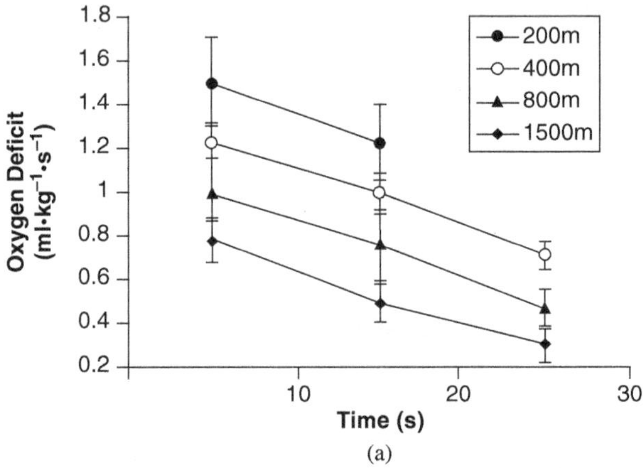

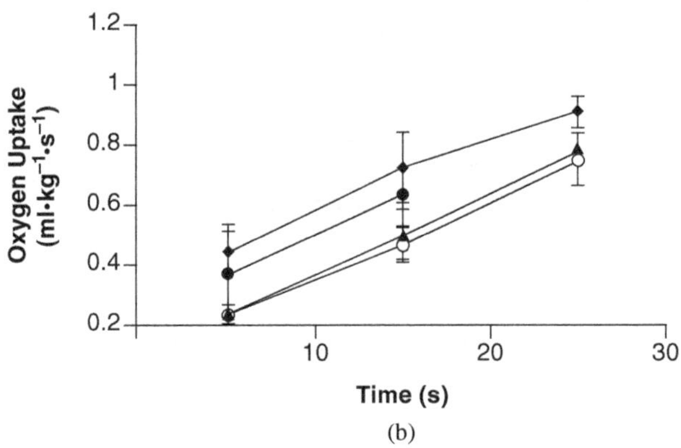

Figure 9. O_2 deficit (A) and O_2 uptake (B) during the initial 30 s of exercise for 200, 400, 800, and 1,500 m. Data are mean ± SD and are presented in 10-s time intervals.

Questions (70 marks)

1. At what percentage of VO_{2max} was your participant running in the treadmill and track runs? Do you think this point was above his anaerobic threshold/lactate threshold? If your participant was exercising above VO_{2max}, how was this possible? (10 marks)
2. At what time in both treadmill and track runs did the aerobic system become the major contributor of ATP? Was there any difference between runs? If yes, why do you think so? (Figure 6) (7 marks)

3. Did the participant reach a steady state of O_2 uptake in both treadmill and track runs? Explain why he/she did or did not. (Figure 7) (6 marks)
4. Were the total aerobic and anaerobic (MAOD) contributions to the total O_2 cost similar for both 1,500-m treadmill and track runs? Why or why not? (Figure 8, 1,500-m run) (10 marks)
5. If the test were conducted over 3000 m, would you expect the MAOD to be higher than that seen in these tests? Why? (7 marks)
6. How do you use the knowledge learnt in understanding MAOD for practical applications in schools? (5 marks)
7. How do you use the knowledge learnt in understanding the relative energy system contributions for coaching and training purposes? Explain. (5 marks)
8. Is there a limitation for the individual's anaerobic system in the various events (400-m, 800-m, and 1,500-m, Figure 8) as researched by Spencer, Gastin, and Payne (1996) with elite athletes? Do you think your participant in this laboratory session has similar aerobic and anaerobic contributions? Is there a limitation for your participant? (10 marks)
9. Plot a line graph of O_2 uptake and O_2 deficit for the first 30 seconds for both track and treadmill 1,500-m runs (Figure 9). Explain the relationship between time and O_2 in both graphs. (10 marks)

Laboratory Session 8

Excess Post-Exercise Oxygen Consumption (EPOC)

Effects of Intensity and Duration on Exercise and EPOC (Recovery)

Recovery from Exercise (Intense or Light)

- After exercise has stopped, there is a sustained elevated VO_2 (EPOC) (Laboratory Session 7 has some details on EPOC)
- Greater glycogen synthesis occurs during a passive recovery
- Maximal glycogen synthesis requires carbohydrate ingestion (0.7 g/kg/hr) (Robergs, 1991)
- Muscle damage caused by exercise slows glycogen synthesis
- An active recovery prevents glycogen synthesis in slow-twitch fibres (metabolic load during running requires energy)

At the end of exercise (intense or light), muscle metabolism differs from steady state conditions (Figures 1 & 2):

- Near maximal blood flow
- Larger increases in glycolytic intermediates
- Larger increase in muscle lactate
- Larger increases in muscle temperature
- Larger increases in catecholamine hormones

(Powers & Howley, 2009)

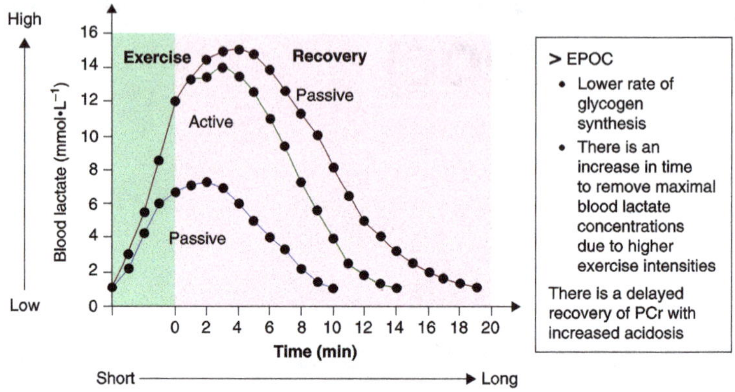

Figure 1. The influence of active and passive recovery after different exercise intensities during excess post-exercise oxygen consumption (EPOC) (adapted from Powers & Howley, 2009).

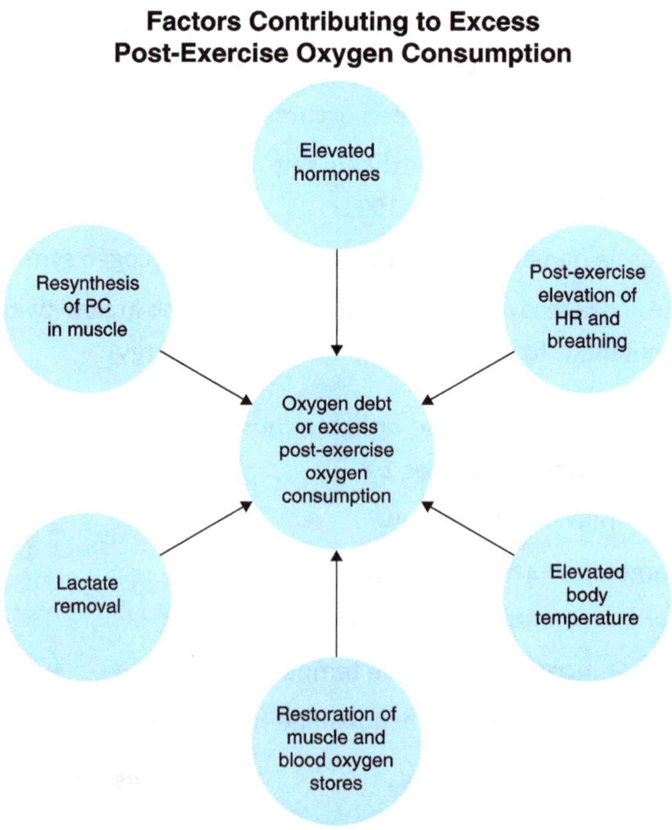

Figure 2. The factors that contribute to excess post-exercise oxygen consumption (EPOC) (adapted from Powers & Howley, 2009).

Metabolic Response to Exercise: Short Duration and High Intensity Exercise

Duration	Intensity	System
10 seconds	High	ATP production (ATP-PC system)
45 seconds	High	ATP production (ATP-PC and anaerobic glycolysis)
3 minutes	High	ATP production (ATP-PC, glycolysis, and aerobic systems)

(See *Applied Physiology of Exercise* Chapters 2, 3, & 4, Balasekaran, Govindaswamy, Lim, Boey, & Ng, 2021)

Metabolic Response to Exercise: Prolonged Exercise

Duration	Intensity	System
10 minutes	Moderate	ATP production (mainly from aerobic systems) Steady state oxygen uptake can be generally maintained
> 1 hour	Moderate	ATP production (mainly from aerobic systems) Steady state oxygen uptake can be generally maintained

(Refer to *Applied Physiology of Exercise* Chapter 4, Balasekaran, Govindaswamy, Lim, Boey, & Ng, 2021)

Metabolic Response to Exercise: Incremental Exercise

- Oxygen uptake increases linearly until VO_{2max} is reached (Figure 3) and there is no increase in VO_2 with increasing work rate
- Physiological factors influencing VO_{2max}:
  - Affects the ability of cardiorespiratory system to deliver oxygen to working muscles (central system)
  - Affects the ability of muscles to utilise the oxygen and produce ATP aerobically (peripheral system)
  - Training at velocity of VO_{2max} is high intensity (Figure 3)

(Refer to *Applied Physiology of Exercise* Chapter 5, Balasekaran, Govindaswamy, Lim, Boey, & Ng, 2021)

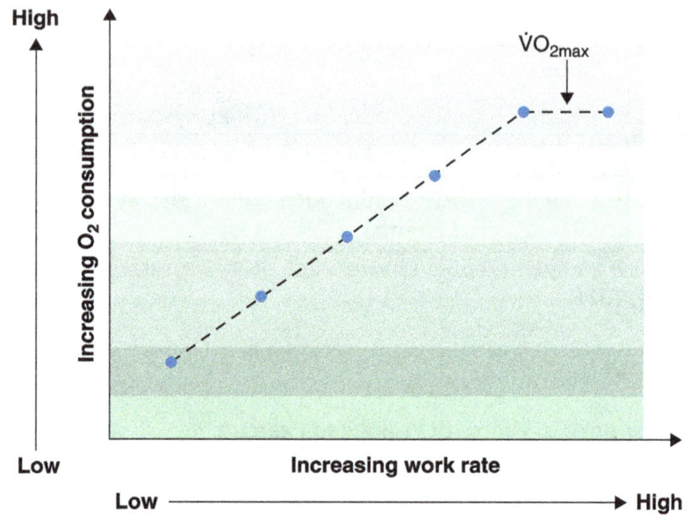

Figure 3. Linear positive regression graph between exercise work rate (watts) and oxygen consumption (VO_2) (adapted from Powers & Howley, 2009).

Questions (40 marks)

1. Why is recovery from intense exercise important for improved sports performance? (8 marks)
2. What would be a better recovery method from intense exercise — active or passive? Why? (6 marks)
3. What are examples of sports or athletic events where individuals do not use appropriate recovery conditions between bouts? (6 marks)
4. What affects EPOC after a hard 2.4-km run on the track? (10 marks)
5. How does duration and intensity of exercise affect EPOC? (10 marks)

Excess Post-Exercise Oxygen Consumption (EPOC) Light Exercise (20 to 30% of HR_{max})

Equipment needed:

1. Treadmill: set at a gradient of 1% to reflect the energy cost of running outdoors

2. Metabolic cart (to measure oxygen consumption; please ensure that it is calibrated)
3. Polar heart rate monitor transmitter
4. Spirometry head gear
5. Nose clip
6. Safety harness
7. Water for participant

Prior to the test, the participant will commence warm-up. He/she will jog an easy or slow pace on the treadmill for 3–5 minutes, followed by simple major muscle group stretches. Allow the participant to rest for 5 minutes. Following which, the participant will run for 10 minutes on the treadmill (light exercise: 20–30% HR_{max} (HR_{max} = 220 – age)). Utilise the oxygen consumption machine, as indicated in previous laboratories, and monitor the participant's heart rate (HR), volume of oxygen (VO_2), volume of carbon dioxide (VCO_2), and respiratory exchange ratio (RER). Once the participant finishes his/her exercise, monitor the physiological variables till they reach the original baseline. Please continue to allow your participant to wear the spirometry head gear connected to the metabolic cart during recovery. The physiological variables would be back to pre-exercise levels. Note the time taken to return to baseline. Draw a similar graph with your participant's data as shown in Figure 4a (VO_2 against exercise time).

Allow the participant to rest for 20 minutes before they commence heavy exercise (70%–80% of HR_{max}).

The participant will run for 10 minutes on the treadmill (heavy exercise). Utilise the oxygen consumption machine, as indicated in previous laboratories, and monitor HR, VO_2, VCO_2, and RER. Once the participant finishes his/her exercise, monitor the physiological variables till they reach the original baseline. Please continue to allow your participant to wear the spirometry head gear connected to the metabolic cart during recovery. The

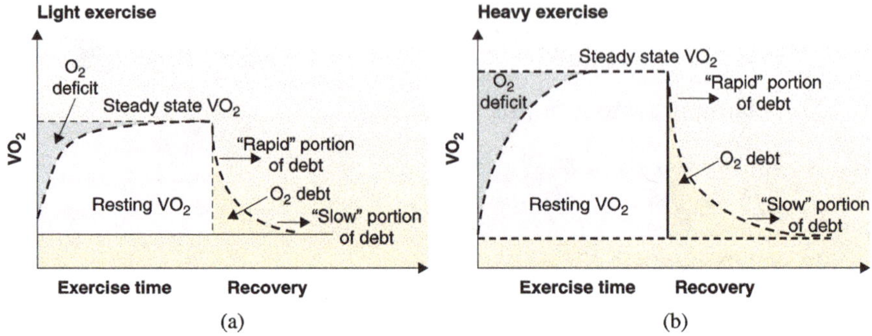

Figure 4. Excess post-exercise oxygen consumption (EPOC) (oxygen debt) comparison between light and heavy exercise (adapted from Powers & Howley, 2009).

physiological variables would be back to pre-exercise levels. Note the time taken to return to baseline. Draw a similar graph with your participant's data as shown in Figure 4b (VO_2 against exercise time).

Table 1. Light intensity exercise data.

Duration (min)	Treadmill Velocity (km·h^{-1})	RER	VO_2 (mL·kg^{-1}·min^{-1})	VO_2 (L·min^{-1})	VCO_2 (L·min^{-1})	Heart Rate (beats·min^{-1})
1						
2						
3						
4						
5						
6						
7						
8						
9						
10						

Table 2. After light intensity exercise — EPOC (recovery) data.

Duration (min)	Treadmill Velocity (km.h^{-1})	RER	VO_2 (mL·kg^{-1}·min^{-1})	VO_2 (L·min^{-1})	VCO_2 (L·min^{-1})	Heart Rate (beats·min^{-1})
1						
2						
3						
4						
5						
6						
7						
8						
9						
10						

*Note: If participant does not return to pre-exercise physiological baseline, you may continue collecting data.

Table 3. Heavy intensity exercise data.

Duration (min)	Treadmill Velocity (km.h^{-1})	RER	VO_2 (mL·kg^{-1}·min^{-1})	VO_2 (L·min^{-1})	VCO_2 (L·min^{-1})	Heart Rate (beats·min^{-1})
1						
2						
3						
4						
5						
6						
7						
8						
9						
10						

Table 4. Heavy intensity exercise — EPOC (recovery) data.

Duration (min)	Treadmill Velocity (km·h^{-1})	RER	VO_2 (mL·kg^{-1}·min^{-1})	VO_2 (L·min^{-1})	VCO_2 (L·min^{-1})	Heart Rate (beats·min^{-1})
1						
2						
3						
4						
5						
6						
7						
8						
9						
10						

*Note: If participant does not return to pre-exercise physiological baseline, you may continue collecting data.

Questions (40 marks)

1. Draw the graphs for light and heavy intensity exercises and EPOC (recovery) as indicated in Figures 4a and 4b. (10 marks)
2. Is there any difference in EPOC between heavy and light exercise? If there is a difference, please explain with physiological facts. (12 marks)
3. Can EPOC measurement be used for anaerobic measurement of exercise performance? Explain why. (10 marks)
4. Does EPOC equate to oxygen deficit? Explain. (8 marks)

Laboratory Session 9

Steady State versus Maximal Exercise

Energy catabolized during exercise is dependent on two factors — diet and intensity of physical activity. As shown in the 'chicken laboratory' (Laboratory Session 3) test, protein diet prior to exercise may induce a longer time to metabolise and may hinder exercise performance. However, intensity of physical activity also determines the type of fuel catabolized. The higher the intensity of the physical activity, the greater need for carbohydrates over fats, as shown in Figure 1.

Exercise intensity is also related to time. As duration increases, the body will favor fat over carbohydrates. In reality, the body will not be able to keep up with a high-intensity exercise for long duration (Figure 2).

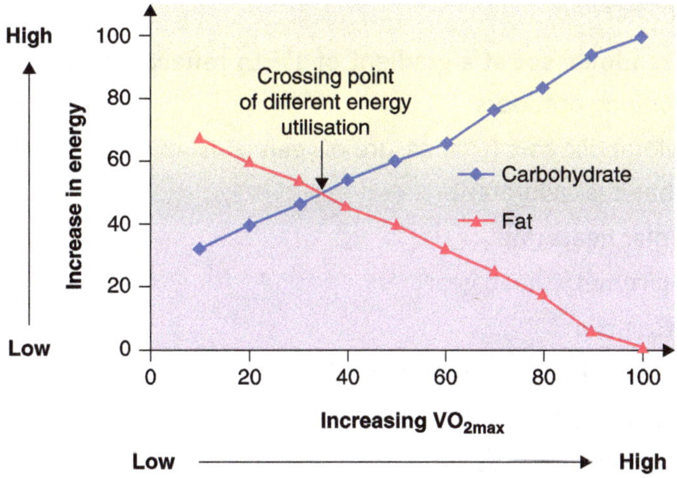

Figure 1. Fuel utilisation as a function of VO_{2max} (adapted from Powers & Howley, 2009).

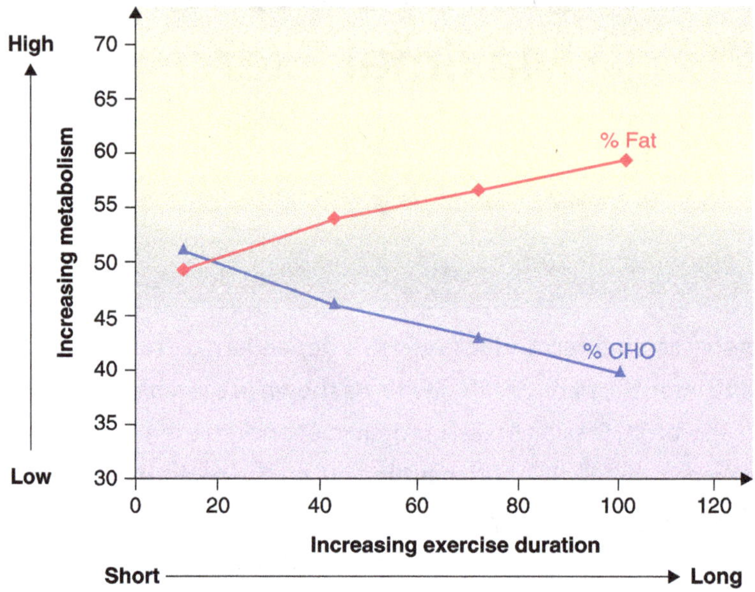

Figure 2. Fuel utilisation as a function of exercise time (adapted from Powers & Howley, 2009)

Fuel Metabolism Laboratory

Purpose: to evaluate the energy substrates used during steady state submaximal and maximal exercise.

Equipment needed:

1. Treadmill: set at a gradient of 1% to reflect the energy cost of running outdoors
2. Metabolic cart (to measure oxygen consumption; please ensure that it is calibrated)
3. Polar heart rate
4. Spirometry head gear
5. Nose clip
6. OMNI RPE scale (Robertson, 2004)
7. Safety harness
8. Water for participant

Procedure:

1) Select 2 participants for evaluation
2) Determine heart rate (HR) obtained every minute
3) Administer graded exercise test on each participant (Table 1)
4) Analyse expired gas at every 30 seconds throughout the test

Table 1. Graded exercise protocol.

Workload	Speed (mph)	Grade (%)	Time (minutes)
1	2.0	0	2
2	2.0	5	2
3	2.0	10	2
4	2.5	15	2
5	3.5	15	2
6	4.5	17	2
7	5.0	17	2
8	5.5	17	2
9	6.0	17	2

Formulas:

Variable	Formula
VO_2 fat ($L{\cdot}min^{-1}$)	VO_2 total ($L{\cdot}min^{-1}$) $\times$ [(1.0 – RER)/0.3]
VO_2 CHO	VO_2 total – VO_2 fat
VO_2 fat/VO_2 total	% VO_2 from fat
VO_2 CHO/VO_2	% VO_2 from CHO
kcal/min	VO_2 ($L{\cdot}min^{-1}$) $\times$ kcal/L
3500 kcal	1 pound of fat
Minute to expend 3500 kcal	3500 kcal/pound $\times$ 1/ kcal/min

*RER refers to respiratory exchange ratio. VO_2 refers to oxygen uptake. VCO_2 refers to carbon dioxide uptake.

Results can be tabulated in Tables 2 and 3.

Table 2. Participant 1 data.

Stages	Time	Treadmill Speed ($km \cdot h^{-1}$)	VCO_2 ($L \cdot min^{-1}$)	VO_2 ($L \cdot min^{-1}$)	VO_2 ($mL \cdot kg^{-1} \cdot min^{-1}$)	RER	VE ($L \cdot min^{-1}$)	Heart Rate ($beats \cdot min^{-1}$)	OMNI RPE
1									
2									
3									
4									
5									
6									
7									
8									
9									
10									
11									
12									
13									
14									
15									

(Continued)

Table 2. *(Continued)*

Stages	Time	Treadmill Speed ($km \cdot h^{-1}$)	VCO_2 ($L \cdot min^{-1}$)	VO_2 ($L \cdot min^{-1}$)	VO_2 ($mL \cdot kg^{-1} \cdot min^{-1}$)	RER	VE ($L \cdot min^{-1}$)	Heart Rate ($beats \cdot min^{-1}$)	OMNI RPE
16									
17									
18									
19									
20									
21									
22									
23									
24									
25									
26									
27									
28									
29									
30									

Table 3. Participant 2 data.

Stages	Time	Treadmill Speed ($km{\cdot}h^{-1}$)	VCO_2 ($L{\cdot}min^{-1}$)	VO_2 ($L{\cdot}min^{-1}$)	VO_2 ($mL{\cdot}kg^{-1}{\cdot}min^{-1}$)	RER	VE ($L{\cdot}min^{-1}$)	Heart Rate ($beats{\cdot}min^{-1}$)	OMNI RPE
1									
2									
3									
4									
5									
6									
7									
8									
9									
10									
11									
12									
13									
14									
15									

(*Continued*)

Table 3. *(Continued)*

Stages	Time	Treadmill Speed (km•h^{-1})	VCO_2 (L•min^{-1})	VO_2 (L•min^{-1})	VO_2 (mL•kg^{-1}•min^{-1})	RER	VE (L•min^{-1})	Heart Rate (beats•min^{-1})	OMNI RPE
16									
17									
18									
19									
20									
21									
22									
23									
24									
25									
26									
27									
28									
29									
30									

Results/Discussion (Refer to formulas on the previous pages) (50 marks)

1) Determine the VO_2 associated with fat and carbohydrate metabolism during each workload. (10 marks)
2) Plot HR, RER, fat, and carbohydrate metabolism as a function of VO_2 during exercise. What comparisons can be made between participants? Explain any differences or similarities between participants. (20 marks)
3) How many kcals were being expended aerobically during the last minute of **each** workload? How long would it take to lose a pound of fat while exercising at **each** workload? (20 marks)

Note: All calculations should be based upon the average of the last two 30-second values from **each** workload.

If the RER is greater than 1.00, it means only carbohydrates are being metabolized (i.e. 100% carbohydrates, 0% fats) (Laboratory Session 3 Tables 7 & 8).

Laboratory Session 10

Lactate Threshold and Wingate Anaerobic Test

Laboratory Session 10.1

In this laboratory session, there will be 2 trials performed by 3 different participants on 2 separate days. One participant will perform the lactate threshold session and the other 2 participants (male and female) will perform the Wingate anaerobic test (Refer to *Applied Physiology of Exercise* Chapter 3, Balasekaran, Govindaswamy, Lim, Boey, & Ng, 2021).

Determination of Lactate Threshold

Pre-trial: Determine the descriptive statistics for the participant. You can use the bioelectric impedence analysis (BIA) machine or the dual-energy X-ray absorptiometry (DEXA) machine for body composition.

Trial 10.1: Submaximal Till Exhaustion Protocol — Discontinuous Protocol (Gupta & Balasekaran, 2013; Ali, Balasekaran, Hoon, & Gerald, 2017; Balasekaran et al., 2020)

Equipment needed:

1. Treadmill: set at a gradient of 1% to reflect the energy cost of running outdoors
2. Metabolic cart (to measure oxygen consumption; please ensure that it is calibrated)
3. Polar heart rate monitor transmitter
4. Spirometry head gear
5. Nose clip

6. OMNI RPE scale (Robertson, 2004)
7. Safety harness
8. Water for participant

The participant will perform a discontinuous submaximal till exhaustion protocol in order to determine maximal oxygen uptake and lactate threshold. Prior to the test, the participant will commence warm-up. He/she will jog an easy or slow pace on the treadmill for 3–5 minutes, followed by simple major muscle group stretches. In this session, your participant will perform a series of discontinuous treadmill runs lasting 4 minutes for each stage. The treadmill speed will increase with each stage at 0.5 $km \cdot h^{-1}$, which should range between 6–14 $km \cdot h^{-1}$ depending on the ability of your participant and the start or end speed could be higher. Between each stage, your participant will have 4 minutes recovery. During the recovery stage, the researcher* will take a finger-prick blood sample from the participant to measure blood lactate. Once the blood sample is collected in a small tube (e.g. microvettes etc.), measure blood lactate results immediately using the blood lactate machine (portable blood lactate analysers may not indicate accurate results). Between each stage, your participant will have 4 minutes recovery time. Steady state VO_2 for each workout will be determined around the final 2 minutes of each submaximal run and ratings of perceived exertion (RPE) using the OMNI RPE scale for adults (Robertson, 2004). HR will also be recorded during the final 2 minutes of each stage. This will help develop a relationship between VO_2, lactate, HR, OMNI RPE, and treadmill velocity. The test will terminate when the desired lactate has been determined. Desired lactate can refer to the following:

1. When plotting a speed versus lactate graph, the linear graph shows a breakaway point or abrupt non-linear rise
2. When lactate values rise above 4 mmol•L^{-1} for 2 stages consecutively

After approximately 20 minutes, Balasekaran et al., (2020) protocol will be used to determine VO_{2max}. While recalibrating the metabolic machine, the participant can perform some major muscle group stretches. Once the participant is geared up with the necessary equipment for the metabolic measurement, including HR, start the VO_{2max} test on him/her. Start the treadmill speed at the same speed as the 9th submaximal treadmill run/10 km•h^{-1} (depending on your participant's start speed). Increase the treadmill velocity by 1 km•h^{-1} for 6 minutes (to 15 km•h^{-1}) (depending on your participant's start speed) at every minute. If your participant is still running, increase the treadmill gradient by 2 %•min^{-1} while keeping speed constant until volitional exhaustion of the participant. Encourage your participant to continue running for as long as he/she possibly can. Exercise will continue till volitional exhaustion of the participant (HR ≥ 95% of maximal HR (max HR = 220 – age)) (for VO_{2max} criteria, refer to American College of Sports Medicine Guidelines, 2018; *Applied Physiology of Exercise* Chapter 5, Balasekaran, Govindaswamy, Lim, Boey, & Ng, 2021). Post-lactate will be determined after the test.

*Researcher to wear surgical gloves during test for hygiene purposes. Dispose of laboratory consumables (e.g. finger prick needle, used tissue, etc.) appropriately into the respective disposal bins according to your laboratory's rules and regulations.

Results can be tabulated as follows (Table 1).

Fill in the following Table 1 with your participant's submaximal till exhaustion data (remember to average the data from the final 2 minutes of each exercise stage).

Name of Participant: ____________________

Weight: _______ Height: ________ BMI: ________ Age: ______

Predicted Max HR: __________ Body Fat %: __________

95% of Maximal HR: __________________

Maximum HR: ____________________

Final stage on treadmill: _______________________

Total exercise time on treadmill: _____________________

Use the data in Table 1 to draw the submaximal regression equation as indicated in Laboratory Session 7 Table 3 and Figure 5. Use this submaximal linear relationship to determine some of the values for the laboratory sessions.

In Table 2, record the participant's data from the VO_{2max} test.

Table 1. Data on submaximal exercise.

Stages	Time	Treadmill Velocity ($km{\cdot}h^{-1}$)	Heart Rate ($beats{\cdot}min^{-1}$)	VCO_2 ($L{\cdot}min^{-1}$)	VO_2 ($L{\cdot}min^{-1}$)	VO_2 ($mL{\cdot}kg^{-1}{\cdot}min^{-1}$)	RER	OMNI RPE	VE ($L{\cdot}min^{-1}$)	Lactate ($mmol{\cdot}L^{-1}$)
1										
2										
3										
4										
5										
6										
7										
8										
9										
10										
11										
12										
13										
14										
15										

(*Continued*)

Table 1. *(Continued)*

Stages	Time	Treadmill Velocity ($km \cdot h^{-1}$)	Heart Rate (beats$\cdot min^{-1}$)	VCO_2 ($L \cdot min^{-1}$)	VO_2 ($L \cdot min^{-1}$)	VO_2 ($mL \cdot kg^{-1} \cdot min^{-1}$)	RER	OMNI RPE	VE ($L \cdot min^{-1}$)	Lactate ($mmol \cdot L^{-1}$)
16										
17										
18										
19										
20										
21										
22										
23										
24										
25										
26										
27										
28										
29										
30										

Table 2. Data on the VO_{2max} test.

Stages	Time	Treadmill Velocity ($km \cdot h^{-1}$)	Heart Rate ($beats \cdot min^{-1}$)	VCO_2 ($L \cdot min^{-1}$)	VO_2 ($L \cdot min^{-1}$)	VO_2 ($mL \cdot kg^{-1} \cdot min^{-1}$)	RER	VE ($L \cdot min^{-1}$)	OMNI RPE

Post-blood lactate immediately after $\dot{V}O_{2max}$ test: _____ $mmol \cdot L^{-1}$

Laboratory Requirements:

Tabulate the participant's results from Table 1 into an Excel spreadsheet and draw the lactate responses to running speed. Indicate a visual lactate threshold determination based on the diagram.

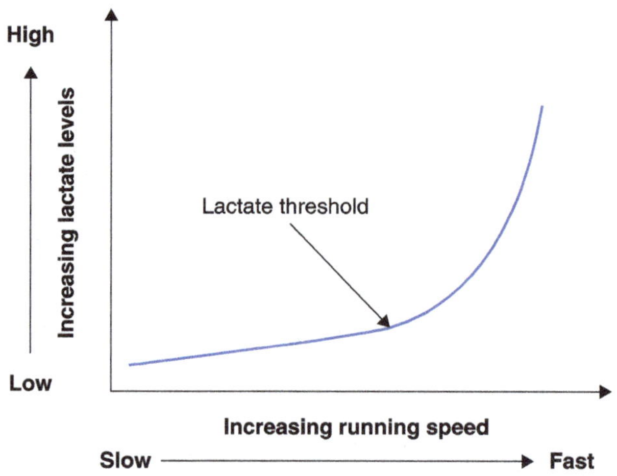

Figure 1. Lactate versus running speed graph.

Questions Section 1 (80 marks)

1. Draw a lactate versus running speed graph (Figure 1) based on the values from your laboratory session and use it to determine the lactate threshold at the critical speed (speed at lactate threshold*): ____________________. (5 marks including diagram)

*Note: lactate threshold in this laboratory manual is used interchangeably with anaerobic threshold. However, some researchers may have different definitions for lactate threshold and anaerobic threshold.

Transfer the data to any statistical software (e.g. SPSS, Excel) to perform simple statistics to determine oxygen uptake and treadmill speed relationship using a linear regression graph (refer to Laboratory Session 7 Table 3 and Figure 5; Appendix A Figure B). Thereafter, determine the lactate threshold using linear regression log-log plot (refer to Figure 2; Appendix A Figure C).

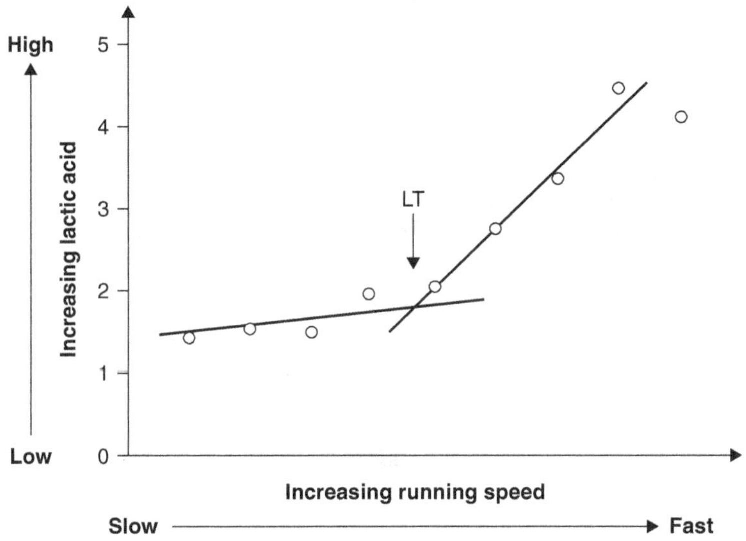

Figure 2. Lactate versus running speed graph with linear regression.

Questions

2. Draw a lactate versus running speed graph based on the values from your laboratory session using linear regression log-log plot and then determine the lactate threshold at critical speed (Figure 2 — critical speed is the speed where lactate threshold occurs). Is this similar to the first determination in question 1? Explain. (Refer to Appendix A) (9 marks including diagram)
3. Calculate the velocity ($km{\cdot}hr^{-1}$) corresponding to 50% and 60% VO_{2max} (refer to Appendix A Figure F). (5 marks)
4. Calculate the heart rate corresponding to 50 and 60% VO_{2max} (refer to Appendix A Figure G). (5 marks).
5. What % of VO_{2max} does the participant's lactate threshold occur? (5 marks)

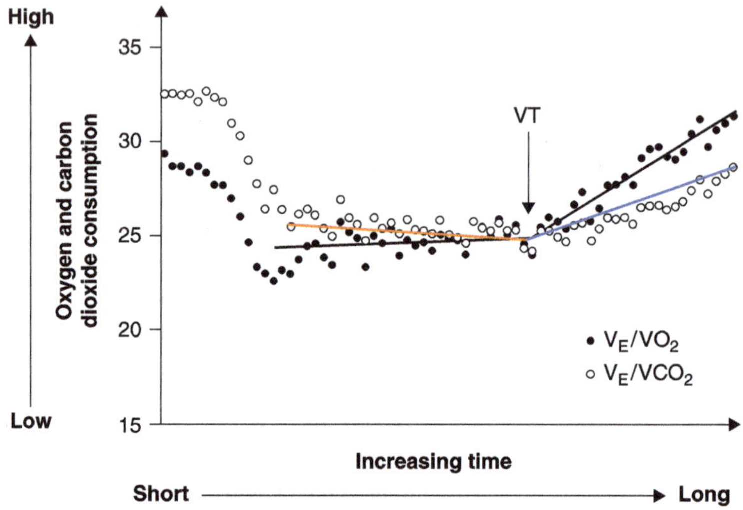

Figure 3. The above figure illustrates the method used for detecting the ventilatory threshold (VT). The VT occurs at the level of exercise intensity at which there is a systematic increase in the ventilatory equivalent for oxygen (VE/VO_2) without a concomitant increase in the ventilatory equivalent for carbon dioxide (VE/VCO_2). Please refer to Appendix A to plot this graph (for individual participants) *Note: ventilatory breakpoint (V_{pt}) is used interchangeably with ventilatory threshold (VT). However, some researchers may have different definitions for ventilatory threshold and ventilatory breakpoint.

Questions

6. Draw the ventilatory threshold (VT) diagram using the data from the laboratory session and determine the point of VT (Figure 3; Appendix A Figure E). Is this point or running speed (critical speed) of VT similar to the lactate threshold determined in Question 2? If so, explain briefly. (10 marks including diagram)
7. Do the lactate threshold and VT occur at the same time? Explain why. (9 marks)
8. Plot changes in ventilation (VE) against the increase of workload/power output/speed and VO_2 against workload/power output/speed to determine VT (Figure 4) (Hint: you can either draw all 3 graphs exactly like Figure 4 or draw VE and VO_2 against workload/power output/speed separately so that the value for VT can be determined. This value can be compared with LT). Does the VT occur at the same time as the one calculated in Question 7? (12 marks)

Questions

9. Determine the HR at lactate threshold using the data from the laboratory session. (Figure 5) (10 marks) (Refer to Appendix A Figure G)
10. a) How will you use HR at lactate threshold for training purposes? (5 marks)

    b) Are there any other ways to use lactate theshold for training purposes other than HR? (5 marks)

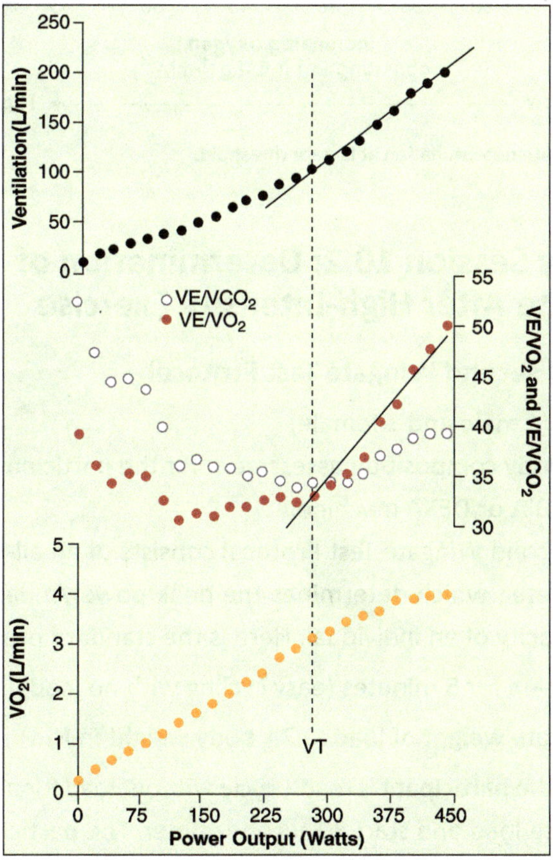

Figure 4. The intensity corresponding to a nonlinear deviation in the increase in ventilation and the point where VT occurs at the level of exercise intensity. There is a systematic increase in the ventilatory equivalent for oxygen (VE/VO_2) without a concomitant increase in the ventilatory equivalent for carbon dioxide (VE/VCO_2) (adapted from Plowman & Smith, 2007).

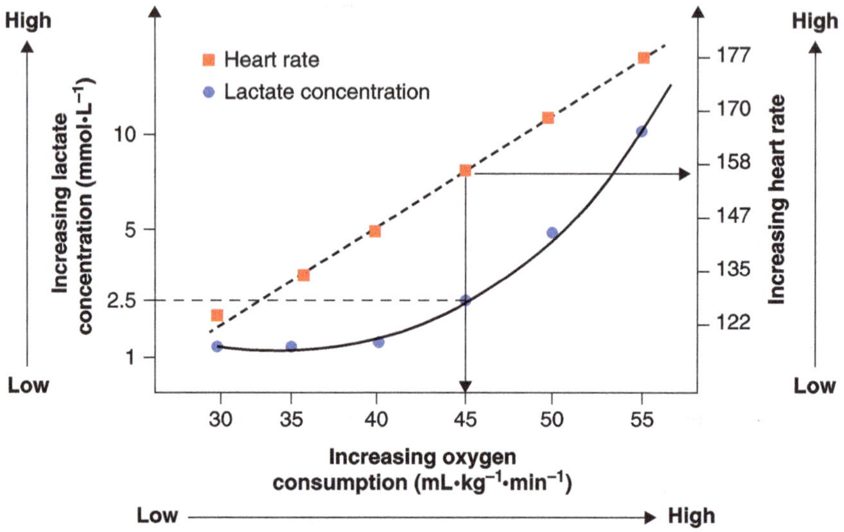

Figure 5. Heart rate determination at lactate threshold.

Laboratory Session 10.2: Determination of Post-Lactate After High-Intensity Exercise

Trial 10.2: 30-second Wingate Test Protocol

2 participants (1 male and 1 female)

You can do a body composition assessment for the participant before the test using the BIA or DEXA machine.

The 30-second Wingate Test Protocol consists of an all-out sprint on a cycle ergometer, which determines the peak power/mean power and anaerobic capacity of an individual. Here is the standard protocol:

1. Warm-up for 5 minutes (easy cycling with no load)
2. Calculate weight of load (0.7 x body weight in Nm)
3. Once the participant is ready, cycle without load for a few seconds. Release load and start 30 seconds timer. The participant to keep cycling and maintain at least 60 rpm throughout the test.
4. The test will terminate at 30 seconds.

5. Take post-blood lactate* via finger-prick method every minute for the next 10 minutes.
6. Repeat steps 1 to 5 for participant 2.

*Researcher to wear surgical gloves during test for hygiene purposes. Dispose of laboratory consumables (e.g. finger prick needle, used tissue, etc.) appropriately into the respective disposal bins according to your laboratory's rules and regulations.

Name of Participant: ____________________

Weight: ________ Height: __________

BMI: __________ Body Fat Percentage (%): __________

Age: ________ Gender: _______

Fill in the following Table 3 with your participant's Wingate test post-lactate data.

Table 3. Data on the Wingate test for participant 1.

Time (min)	Blood Lactate ($mmol{\cdot}L^{-1}$)
1	
2	
3	
4	
5	
6	
7	
8	
9	
10	

Name of Participant: ____________________

Weight: ________ Height: __________

BMI: __________ Body Fat Percentage (%): __________

Age: ________Gender: _______

Fill in the following Table 4 with your participant's Wingate test post-lactate data.

Table 4. Data on the Wingate test for participant 2.

Time (min)	Blood Lactate (mmol•L^{-1})
1	
2	
3	
4	
5	
6	
7	
8	
9	
10	

Questions Section 2 (60 marks)

11. Draw a graph on lactate over time for both participants. (5 marks)
12. Comment on the physiology of lactate responses of both participants from Tables 3 and 4 after a bout of high-intensity exercise. (5 marks)
13. Are there any differences in post-lactate responses after a high-intensity Wingate test between male and female participants? Explain. (8 marks)
14. A 30-second bout is sufficient to cause a lactate overload. What energy systems are being utilised in the Wingate test? Explain. (8 marks)

15. In field settings (on the track), where it is possible to induce the highest lactate possible, what kind of principles will you use to elicit lactate in your participants? Give examples of structured workouts and give physiological rationale behind the workouts. (10 marks)
16. The lactic acid system is usually associated with sports activities such as sprinting, but elite endurance athletes (which includes soccer, tennis, basketball, long distance running, etc.) may also use this system.
    a. Using a sporting example, explain why an endurance athlete may need to utilise the lactic acid energy system. (5 marks)
17. During a 400-m race, an athlete's level of lactic acid can increase tenfold.
    a. What effect can this increase in lactic acid **during** running have on muscle function? (5 marks)
    b. Explain what can happen to the lactic acid produced in muscles **during** exercise. (5 marks)
18. For convenience and ease, how is "maximal oxygen consumption" usually written/expressed? (2 marks)
19. If somebody has a maximum oxygen uptake of 2.5 litres/minute, how would you describe what this means to them in everyday language? (2 marks)
20. What factors influence the size (level — fit or unfit) of an individual's maximum oxygen uptake? (3 marks)
21. Following testing, a male participant weighing 80 kg has a VO_{2max} of 3.21 $L{\cdot}min^{-1}$, and a female participant weighing 50 kg had a VO_{2max} of 2.4 $L{\cdot}min^{-1}$. Relatively speaking, who was the fittest aerobically? Show your workings. (5 marks)

Questions Section 3 (60 marks)

1. Record the following data from the participant in the laboratory session for $\mathbf{VO_{2max}}$.
    a. Age (1 mark)
    b. Weight (1 mark)
    c. VO_2, HR, VCO_2, and RER in the final minute of each stage (make a table with the data) (6 marks)

i. Did the participant reach VO_{2max}? Support your answer using the American College of Sports Medicine criteria (for VO_{2max} criteria, refer to American College of Sports Medicine Guidelines, 2018; *Applied Physiology of Exercise* Chapter 5, Balasekaran, Govindaswamy, Lim, Boey, & Ng, 2021). (5 marks)

ii. If VO_{2max} was reached, what was the post-blood lactate value? (1 mark)

1. What was the VO_{2max} in both relative and absolute terms? (2 marks)
2. What aerobic fitness classification would you give the participant? (Refer to own research or *Applied Physiology of Exercise* Chapter 5, Balasekaran, Govindaswamy, Lim, Boey, & Ng, 2021) (2 marks)

2. What is RER and how is it calculated? (Hint: Refer to Laboratory Session 3; *Applied Physiology of Exercise* Chapter 5, Balasekaran, Govindaswamy, Lim, Boey, & Ng, 2021) (2 marks)
3. Describe the participant's RER during the whole **VO_{2max} test**. (filling Table 5 may help you answer the question) (5 marks)

Table 5. Data on RER.

Time (min)	RER
1	
2	
3	
4	
5	
6	
7	
8	
9	
10	

4. What factors need to be considered before, during, and after the VO_{2max} test to maintain the health and well-being of the participant? (4 marks)
5. What is the progressive aerobic cardiovascular endurance run (PACER) test? Is it a good indicator of fitness level? Explain. (3 marks)
6. Which maximal oxygen test would you administer for students? Give reasons for your choice. (5 marks)
7. What would you do to increase your values on the PACER test? Elaborate your answer with proper training and energy systems principles. (7 marks)
8. The right ventricle of the heart is less muscular than the left ventricle.
   a. With the right ventricle being smaller, explain the effects this has on the blood going to the lungs. (3 marks)
   b. What is the effect or advantage of the largeness of the left ventricle? (3 marks)
9. Complete Table 6 indicating which energy system would be predominant in supplying the required energy. (10 marks)

Table 6. Predominance of energy system.

Exercise	Duration	Predominant Energy System
100-m sprint	10 secs	
Marathon	3 hr 45 mins	
100-m swim	64 secs	
Gymnastics vault	4 secs	
400 m	60 secs	
1,500 m	5 min 09 secs	
Tennis serve	6 secs	
Resting	> 30 mins	
5-km jog	30 mins	
Boxing round	2 mins	

Laboratory Session 11

Field Implementation for Lactate Threshold and Aerobic/Anaerobic Interval Training

Aerobic Conditioning

- For health-related components 40% to 89% of heart rate (HR) reserve method or Karvonen formula (ACSM, 2015). Note: Look at terminology below as low intensity can be used for fat loss.
- For performance — 80% to 89% of heart rate (HR) reserve method or Karvonen formula (ACSM, 2015). Note: Look at terminology below as coaches/sports practitioners can use a wider range from 60% onwards.
- Used predominantly in the general preparation or conditioning period, however it still needs to be used in the specific training period and competition training periods. If not, deconditioning in cardiovascular fitness will take place.

Phases of Training

Transition (4 weeks)
Early background (12 weeks)
Pre-season preparation (12 weeks)
Early season (12 weeks)
Early peak season (8 weeks)
Peak (6 weeks)
Pre-competition prep (9 days before competition)

Note: The above terminology may differ as some coaches/practitioners may interchange terms such as "general phase", "specific phase", or "competition phase". Number of weeks for each phase may differ depending on the type of athletes the coach/practitioner is working with.

Terminology for Interval Training

Low-Intensity Interval Training (LIIT): utilisation of fat substrate (30%–39% of HRR, ACSM, 2015)

Moderate-Intensity Interval Training (MIIT): utilisation of carbohydrate substrate (40%–59% of HRR, ACSM, 2015)

Vigorous-Intensity Interval Training (VIIT): utilisation of carbohydrate substrate (60%–89% of HRR, ACSM, 2015)

High-Intensity Interval Training (HIIT): utilisation of carbohydrate substrate with lactate production (maximal HR)
High-Intensity Interval Training (HIIT) with very short duration (3–15 secs): utilisation of phosphocreatine substrate with no lactate production (maximal HR)

*HRR: Heart Rate Reserve, HR: Heart Rate; Karvonen formula: Target Heart Rate = [(max HR – resting HR) × % Intensity] + resting HR, inclusion of resting HR in this formula makes it more accurate than maximal HR formula (220 – age). Resting HR changes as one gets fitter.

You can link the following various interval trainings to the terminology mentioned and remember that it might be erroneously used due to the inadequate knowledge of exercise prescription of intensity. For example, HIIT is an all-out intensity level where maximal effort is needed and HR is at maximal level with high level lactate production. On the contrary, most individuals erroneously exercise at the intensity between MIIT and HIIT and term this exercise as HIIT. Hence, this intensity is not HIIT as MIIT is 40%–59% of HRR.

Interval Training

Interval training/fartlek training can be used for any aerobic endurance sport (soccer, basketball, hockey, rugby, etc., not exclusively for track and field only).

You may not have access to lab facilities, but you have the laboratory experience and knowledge in lactate physiology (Laboratory Session 10; *Applied Physiology of Exercise* Chapter 3, Balasekaran, Govindaswamy, Lim, Boey, & Ng, 2021). Using this knowledge design, write out an anaerobic and aerobic interval training session for yourself or group (this training programme can be used for adults, adolescents, or children. Repetitions, intensity, and recovery may differ). Your description should include each of the following points:

- The objective of the session, e.g. to develop maximum speed or speed endurance or maximum oxygen uptake
- The contribution of the session to the overall training program
- The number of intervals
- The distance of the intervals
- The target time(s) for the intervals
- The mode of recovery
- The duration of recovery
- The training site/surface
- The expected heart rate response i.e. both the pattern of the response and the actual values you expect to record

While designing your program, bear in mind the following:

- For quality (intensity), i.e. anaerobic capacity, emphasise fast running with a low number of repetitions and long recoveries
- For quantity (volume), i.e. aerobic capacity, emphasise low speeds with a high number of repetitions and short recoveries
- For specificity, the running time should not be too much faster or too much slower than race pace
- The session must be specific to the event you are training for, e.g. a 100-m runner would not run 20 × 400 m and a marathon runner would not run 10 × 60 m (they can still do the workout depending

on the season and what the objective of their training session is, e.g. speed or endurance. Please note that the training volume of 20 x 400 m for endurance training is too much for a sprinter). As a guide, the total distance covered in the session should be about twice the race distance, but this will not hold true for all events (overload training principle).

Come prepared to perform your session next lesson on the track! Refer to 11.1 to 11.3 and complete Tables 2, 3, and 4 before the lesson.

Some Pointers on the Type, Repetitions, Intensity, and Volume of Interval Training

Fartlek training (informal/unstructured interval training): can be performed on different terrains—hilly regions, flat terrains, or cross-country terrain. Different terrains will elicit different intensities and volume. For example, running on a hilly region can be classified as strength endurance workout (Balasekaran, 2002).

Instructions to participants has to be clear as the Fartlek training can be aerobic or anaerobic, since the rest, volume, and intensity can be manipulated to elicit different intensities (Balasekaran, 2002).

Interval training (formal/structured training) can be in the form of: repetitions of 100 m, 200 m, 400 m, 600 m, and 800 m. In some instances, enhanced endurance training will also include 1,000 m, 1,200 m, 3,000 m, and 5,000 m.

Active recovery is recommended (usually a jog). The recovery jog increases as repetition distance increases (depends on the fitness of the athlete).

The number of intervals decreases as repetition distance increases. For e.g.

- 12 × 400 m (50-m jog)
- 6 × 800 m (100-m jog)

- 5 × 1,200 m (200-m jog)
- 4 × 1,600 m (200-m jog)

The jog recovery depends on the intensity and fitness of the individual.

Fast Interval Training

Fast interval training (HR > 180 beats•min^{-1} for sub-elite and children. Maximal effort can be used for elite and can be termed as HIIT as it is all-out intensity)

- Speed or anaerobic runs (used in the later part of early season and early peak season)
- The recovery is longer, but the repetition is low at a higher intensity
- 10 × 100 m (add 1.5 seconds to best time in 100 m)
- 3–5 × 200 m (add 3 to 5 seconds to the best time in 200 m)
- 2 × (5 × 400 m) (subtract 1 to 4 seconds from the 400-m pace for best time in 10,000-m or 5,000-m race)
- 2–4 × 600 m (subtract 2 seconds from best time of race)
- 4 or 5 × 800 m (add 2 to 4 seconds to best time in race (recovery time in between sets or repetitions and intensity depend on the fitness of the individual. For more information, refer to *Applied Physiology of Exercise* Chapter 3, Balasekaran, Govindaswamy, Lim, Boey, & Ng, 2021)

Slow Interval Training

Aerobic endurance: used in early background, pre-season preparation. Decreases in early season. The recovery is shortened, but the repetition is high at a lower intensity.

Intensity/pace for intervals:

- Add 4 seconds to best 100 m
- Add 6 seconds to best 200 m
- Add 4 seconds to 400 m for best 10,000 m or 5,000 m time
- Add 6 seconds to 600 m
- Add 8 seconds to 800 m

Recovery time in between sets or repetitions and intensity depends on the fitness of the individual. For more information, refer to *Applied Physiology of Exercise* Chapter 4, Balasekaran, Govindaswamy, Lim, Boey, and Ng, 2021.

Or you can determine pace/intensity by

1) Determining intensity: the highest possible intensity (100%) is taken as the best performance by the individual athlete in a given practice. For example, if an athlete's best 200 m is 22.7 seconds, this would represent 100% intensity. If an athlete runs 200 m repetitions in training at 90% intensity, this would require a time of 25.3 seconds. Each run can be referred to as submaximum training (Table 1). This training will equate to an anaerobic system training (intensity depends on the fitness of the individual).

Table 1. Scale of intensity for training.

	Spheres of Intensity		
↑	**Scale of Intensity**	**% Maximum Intensity**	↑
Anaerobic	Maximum	95–100	High
	Sub-maximum	85–94	
	High	75 –84	
	Medium	65–74	
	Light	50–64	
Aerobic	Low	30–49	Low

2) Determine intensity by knowing the goal of the athlete. For example, if the athlete wants to run 3,000 m (a school soccer player might cover this distance in a game) in 11.00 minutes, he has to run 400 m in 88 seconds. Thus, he does 16 × 400 m at this 88-second pace with a 100-m jog. This training will equate to an aerobic system training (intensity depends on the fitness of the individual).

Note: Do take note of the individual athlete's fitness as he/she may not be able to attempt the 16 x 400 m due to his/her fitness. May need to start with 5 or 6 x 400 m. This is up to the coach's/practitioner's discretion to utilise training principles of progression.

Different Energy Systems Training Programmes

Below are various energy systems training programmes that you can plan for an individual or for yourself. Fill up Tables 2, 3, and 4.

Laboratory Session 11.1: ATP-PC System Training (10 marks)

Objective of session: ______________________________

Training season (refer to phases of training): ______________________

Physiological adaptations* for use of this training during this phase: ______________________________

Table 2. Training programme for ATP-PC system.

Distance (m)	Number of Repetitions (sets)	Rest (minutes/ distance)	Remarks (if any)

Laboratory Session 11.2: Anaerobic System Training (10 marks)

Objective of session: ____________________

Training season (refer to phases of training): ____________________

Physiological adaptations* for use of this training during this phase:

____

Table 3. Training programme for anaerobic system.

Distance (m)	Number of Repetitions (sets)	Rest (minutes/ distance)	Remarks (if any)

Laboratory Session 11.3: Aerobic System Training (10 marks)

Objective of session: ______________________________

Training season (refer to phases of training): ______________________

Physiological adaptations* for use of this training during this phase: ______________________________

Table 4. Training programme for aerobic system.

Distance (m)	Number of Repetitions (sets)	Rest (minutes/ distance)	Remarks (if any)

*Refer to *Applied Physiology of Exercise* Chapter 7, Balasekaran, Govindaswamy, Lim, Boey, and Ng, 2021.

Lactate Threshold Training (Chapter 10) (80 marks)

Determine the participant's lactate threshold at the critical speed in the previous submaximal exercise Laboratory Session 10 and develop an interval training session for the participant at critical speed. (12 marks)

Questions

1. What is the rest in between an anaerobic interval training? Explain the physiological rationale. (8 marks)
2. What is the rest in between an aerobic interval training? Explain the physiological rationale. (8 marks)
3. If you do not have access to the laboratory to determine lactate threshold, how would you determine an aerobic interval pace for your participant at lactate threshold? Give examples and explain. (10 marks)
4. Give an example of an ATP-PC anaerobic interval training. How long a rest would you give in between intervals for full recovery? Explain the physiological rationale. (12 marks)

Laboratory Session 12

Prediction of Running Performances — Running Energy Reserve Index (RERI)

Energy Systems

There are 2 types of energy systems, aerobic and anaerobic, that contribute significantly to different types of running. It is an established fact that even though all energy systems are treated as separate entities, they overlap each other during most activities (Baechle & Earle, 2000). It means that even if the aerobic and anaerobic energy systems dominate in slower and faster activities, respectively, some expenditure of anaerobic energy during slower activities and aerobic energy during faster activities has been observed (Verheijen, 1998). No single energy system is solely responsible for energy contribution in any activity.

Although all of these 3 energy systems fuel running and other activities, each system dominates according to the duration, intensity, and type of activities. Anaerobic energy (ATP-PC and anaerobic glycolysis) dominates in short-duration, high-intensity work or maximal exercises which last up to 2 minutes. Specifically, the ATP-PC system dominates the energy contribution in sprint running for 4–6 seconds (Janssen, 1994), maximally up to approximately 10 seconds (Spriet, 1995), while activities above 2 minutes in duration are dominated by aerobic energy contribution. Brown, Miller, and Eason (2006) stated that the phosphagen system is the predominant energy system for maximal activities lasting 15 seconds or less. Thus, taking it as a range from 3–15 seconds is accurate. (Refer to *Applied Physiology of Exercise* Chapters 1, 2, 3, & 4, Balasekaran, Govindaswamy, Lim, Boey, & Ng, 2021).

Mathematical Models to Predict Running Energy

Considering limitations of various physiological methods to measure anaerobic energy, researchers developed various mathematical models, including hundreds of empirical and theoretical running concepts, to measure running energetics. However, most of these models have focused on the prediction of middle and long-distance running performance, which are highly accurate and established (Peronnet & Thibault, 1989; Joyner, 1991; Di Prampero et al., 1993). However, very few models have been developed to predict high-speed performance (Volkov & Lapin, 1979; Van Ingen Schenau et al., 1991; Ward-Smith, 1999; Arsac & Locatelli, 2002; Bundle et al., 2003; Di Prampero, 2003; Di Prampero et al., 2005) and most of them are not established. Most of these experimental and theoretical models are based on the speed-duration relationship originating from Hill (1925, 1950).

Relation Between Distance and Speed

The speed of sprinters is 2 to 3 times higher than the speed of long-distance runners (maintained for several hours) (Bundle et al., 2003, Mougios et al., 2006). However, maximal speed declines very rapidly as the duration of the activity increases. Sprinters accelerate up to their maximal speed and decelerate as fast as possible with increasing distance (Figure 1;

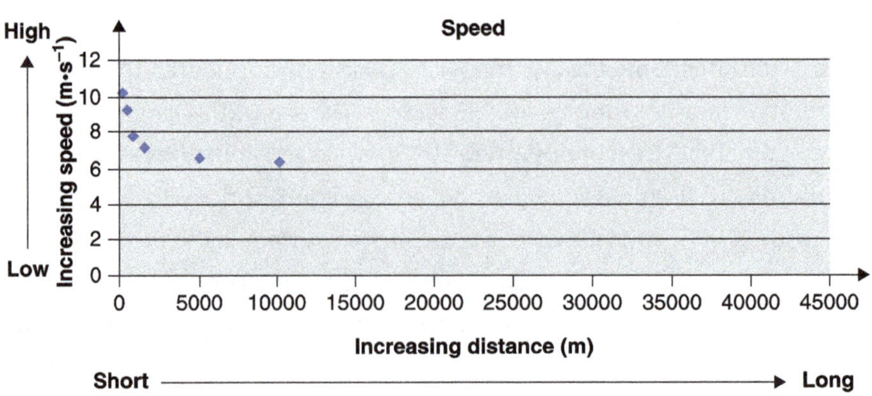

Figure 1. The relationship between speed and distance on the basis of various world records (Hawley et al., 2000).

Note: Data points are available until 10,000 m. Thereafter, speed decreases even more but sustained over a longer distance.

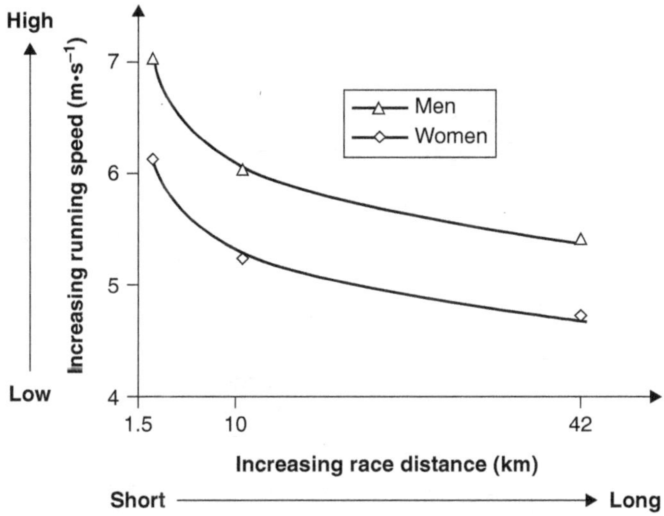

Figure 2. Relationship between run duration and distances among males and females (Sparling et al., 1998) and between run distances and speeds (Sparling et al., 1998). Similarly, Bundle et al. (2003) also determined the running speed at various short durations on the treadmill and observed the same speed duration curves.

Hawley et al., 2000). On the contrary, middle and long-distance running involves steady state VO_2 and speed is maintained for a period of time. If speeds at distances starting from 100 m to 100 km are plotted together, the pattern of speed decrement with increase in duration can be observed. Figure 1 (Hawley et al., 2000) shows that the speed at shorter distance deceases rapidly in a negative exponential manner while speed at long distances declines moderately as duration increases (Bundle et al., 2003).

The same relationship was observed when Sparling et al. (1998) generated a relationship among various running distances ranging from 1,500 m, 10 km, to marathon (42.195 km) and their corresponding speeds ranged from 7.01, 6.01, to 5.39 $m{\cdot}s^{-1}$ for men and 6.12, 5.23, 4.71 $m{\cdot}s^{-1}$ for women, respectively (Figure 2; Sparling et al., 1998). There was no difference obtained between both genders in terms of negative exponential relationship.

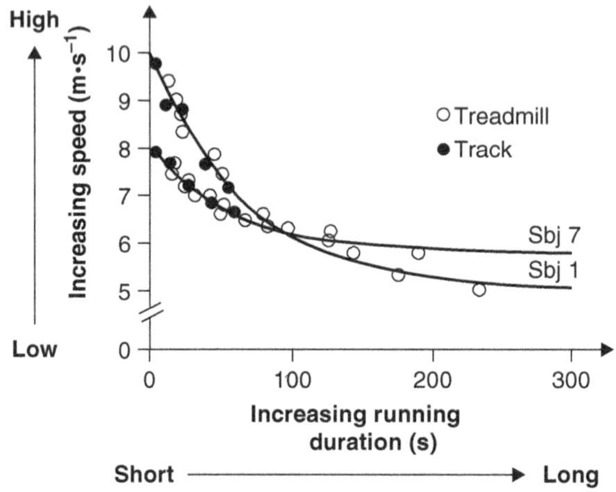

Figure 3. Negative exponential relationship between speed and run duration on the track and treadmill (Bundle et al., 2003).

Curvilinear Relationship Between Speed and Short Distance

It has been found that the relationship between speed and duration/distance rapidly declines in short-distance races (shown in Figures 3 & 4) (Bundle et al., 2003). It has also been observed that this rapid decline in speed is observed up to a duration of 180 seconds. Bundle et al. successfully verified this relationship in various short-distance running ranging from 100–400 m on the track, and 3 to 300 seconds running on the treadmill (Figure 3; Bundle et al., 2003).

The reason behind the observed rapid decline of speed with short distances may be due to the higher involvement of anaerobic sources in comparison to aerobic sources during short-distance running. Bundle et al. (2003) measured the amount of anaerobic energy consumption during running and observed the same negative exponential decline in the rate of consumption of anaerobic energy with increasing distance/duration, as seen in the speed-distance relationship. Ramsbottom et al. (1994) also determined the anaerobic contribution in short-distance running ranging from 100–800 m and observed a high anaerobic contribution in these distances. The declining curve of ATP-PC energy sources matches fairly with the declining curve of speed with various increases in duration or distance

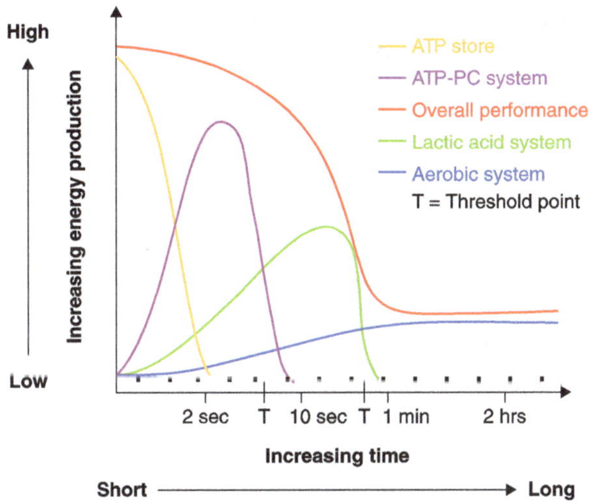

Figure 4. Interaction of various energy systems during running (adapted from Mackenzie, 1997).

(Figure 4; Mackenzie, 1997). Energy contribution from ATP-PC reaches to its maximum in 5–6 seconds (other authors have different timings — 4–6 seconds (Janssen, 1994) or below 15 seconds (Brown, Miller, & Eason, 2006)) and then it follows a negative exponential curve with increasing time (Figure 4; Mackenzie, 1997).

Linear Relationship Between Speed and Long Distances

Hawley et al. (2000) and Sparling et al. (1998) (Figures 1 & 2) showed a linear relationship between longer distances and their corresponding speeds. In middle-distance running, distances such as 2,000 m and 3,000 m, the runners run at a very sustained velocity which is as close as their velocity at VO_{2max} (vVO_{2max}) (Lacour et al., 1990; Billat, 2001). The use of anaerobic energy sources declines, and the rate of aerobic energy contribution reaches its maximal in middle and long-distance running (Table 1). Bundle et al. (2003) also measured the number of aerobic sources in various distances and observed that the aerobic energy contribution reaches its maximal and gets steady with further increase in distance or duration. Table 1 shows that the aerobic contribution in 400 m (35%) increases by

Table 1. Aerobic and anaerobic energy contribution in various running distances (adapted from Hawley et al., 2000).

			Contribution of Energy Systems (%)	
Distance (m)	**Time (s)**	**Speed (m•s^{-1})**	**Anaerobic**	**Aerobic**
100	9.77	10.24	95	5
200	19.32	10.35	85	15
400	43.18	9.26	65	35
800	101.11	7.91	42	58
1,500	206.00	7.28	24	76
5,000	759.36	6.58	7	93
10,000	1,582.75	6.32	4	96
42,195	7,538.00	5.60	1	99

30% as compared to that of 100 m (5%), but only a 6% increase in aerobic contribution is observed when the distance is increased from 5000 m to marathon distance. Similarly, the anaerobic contribution also decreases quickly with increasing distance (Hawley et al., 2000; Duffield & Dawson, 2003).

Anaerobic Speed Reserve

Most of the models mentioned above have been proven to be accurate, but all of them are meant for estimating endurance performance, and not sprinting or high-speed performance. Therefore, keeping in mind the need for developing pragmatic models to predict sprint performance using experimental data, Bundle et al. (2003) developed a model for predicting high-speed performance based on the direct measurement of maximal aerobic and anaerobic power attributed to maximal anaerobic and aerobic speed. Bundle et al. (2003) used the ratio of aerobic power and anaerobic power at various distances (short to long) as this ratio has been observed to be directly related to a decline in anaerobic performance (Di Prampero et al., 1993; Weyand et al., 1999; Savaglio & Carbone, 2000; Bundle et al.,

2003). This relationship has already been verified in sprinting (Van Ingen Schenau et al., 1991), middle-distance running performance (Di Prampero et al., 1993), and in some other activities (Di Prampero, 2003). Using a similar concept in Bundle et al. (2003), maximal aerobic speed (MAS; V_{AerMax} represents maximal aerobic energy) was subtracted from maximal anaerobic speed (V_{AnaMax}), which represents the peak anaerobic energy. The observed value was termed as 'anaerobic speed reserve (AnSR = $V_{AnaMax} - V_{AerMax}$)'. The time constant was derived, which accurately (within an average of 2.5%; $R^2 = 0.94$) predicted the high-speed performance starting from 3 seconds, sprinting up to 240 seconds, using the equation based on the negative exponential relationship between speed and time to exhaustion ($Spd(t) = Spd_{aer} + (Spd_{an} - Spd_{aer}) \times e^{(-k \times t)}$). Bundle et al. (2003) used maximal speeds to represent maximal energy (anaerobic and aerobic). Speed at VO_{2max} and maximal speed at which maximal anaerobic energy obtained within a very short duration (3 seconds or 55 m; Bundle et al., 2003) were termed as maximal aerobic speed and maximal anaerobic speed attributed to maximal aerobic and anaerobic power, respectively.

Maximal Speed Representative of Peak Anaerobic Energy

Many researchers used two terms interchangeably: 'most anaerobic speed' and 'maximal speed' measured during very short-distance running at supramaximal intensity (Bundle et al., 2003). The most anaerobic speed is the fastest speed at which maximal contribution obtained through anaerobic energy sources, or at which maximal anaerobic power is obtained (Bundle et al., 2003; Weyand & Bundle, 2005). The rate of anaerobic energy release during supramaximal exercises reaches its peak in the first few seconds, specifically reaching its maximal rate in 4–6 seconds of supramaximal exercise (Janssen, 1994) in which energy is mainly produced through the ATP-PC system. A poor correlation between sprinting performance and aerobic power with detection of low aerobic power among mammalian sprinters has proven that the maximal anaerobic speed or sprinting speed is mainly supported by anaerobic sources of energy. There was also a high correlation obtained between maximal velocity and muscular power or peak energy expenditure. The concept of maximal speed in maximal exercise has been used in many studies to determine peak anaerobic energy (Bundle et al., 2003; Conley, 2003). Maximal anaerobic speed is usually 2 to 3 times faster than the speed which can be maintained for several hours among world-class runners (according to world records in long distance running and sprinting) (Hawley et al., 2000). In a few studies, the maximal speed (8.7 ± 0.4 $m \cdot s^{-1}$) was found to be 1.7 times faster than the corresponding maximal aerobic speed (5.3 ± 0.1 $m \cdot s^{-1}$) among trained collegiate and post-collegiate runners (Bundle et al., 2003).

Aerobic and Anaerobic Energy Contribution at Supramaximal Speed

Determination of maximal speed is necessary to measure the amount of peak anaerobic energy sources used at this speed. It has been well known that the maximal power attained at maximal velocity is anaerobic but a small fraction of energy during these short-term, high-intensity activities is also supplied through aerobic sources (Hawley et al., 2000). The amount of aerobic contribution during supramaximal exercise was calculated in previous years but it was found to be very low (Astrand, 2003). Arsac and Locatelli (2002) calculated the amount of aerobic and anaerobic energy sources consumed in a 100-m sprint and it was observed that 95% of total energy was consumed from anaerobic sources and the remaining 5% of energy was from aerobic sources. Similarly, Péronnet and Thibault (1989) predicted the sprint energetics of 100-m sprint running and observed 94% energy contribution from anaerobic sources while 6% energy contribution from aerobic sources. Additionally, aerobic power is not correlated with sprinting performance (Svedenhag & Sjodin, 1984; Weyand et al., 1994). Therefore, the amount of oxidative energy used in maximal sprinting is foreseeable and can be ignored.

Maximal Aerobic Speed Representative of Peak Aerobic Energy

Speed at VO_{2max} has been used as maximal aerobic speed (MAS) in many studies (Di Prampero et al., 1986; Lacour et al., 1990, 1991; Billat & Lopes, 2006). This is the minimum speed which elicits maximum contribution from aerobic sources (Di Prampero et al., 1986; Lacour et al., 1990, 1991; Billat & Lopes, 2006). Speed at VO_{2max} has also been termed as 'maximal aerobic speed' by many researchers (Renoux et al., 1999, 2000; Billat & Lopes, 2006). All the muscle fibres reach their maximal aerobic consumption capacity at MAS.

Running Energy Reserve Index Model to Measure Anaerobic Performance

The development of the running energy reserve index (RERI) is based on the determination of maximal anaerobic and maximal aerobic energy contribution based on running speeds (Gupta & Balasekaran, 2013). For this purpose, it is significant to determine the maximal anaerobic speed and maximal aerobic speed.

Equipment needed:

1. Treadmill: set at a gradient of 1% to reflect the energy cost of running outdoors
2. Metabolic cart (to measure oxygen consumption; please ensure that it is calibrated)
3. Polar heart rate monitor transmitter
4. Spirometry head gear
5. Nose clip
6. OMNI RPE scale (Robertson, 2004)
7. Safety harness
8. Water for participant

Questions for Lab Report (160 marks)

Interpret the results of each test below for the participant so that he/she can improve his fitness and performance.

Name of Participant: ________________ Age: ____________

Description or background of participant:

Weight: ______kg Height: ______ m BMI: ________ $kg{\cdot}m^{-2}$

Note: Interpretations mean discuss and compare with norms where possible on participant's physiological variables. If no norms are available, compare with elite or general population results in the literature.

Interpretations of participant's BMI:

(3 marks)

Dual-energy X-ray absorptiometry (DEXA) if available. Bioelectric impedence analysis (BIA) is the alternate method. If both methods are available, interpret the results for both methods.

Body fat: ____%

Interpretations of participant's body fat %:

(4 marks)

Determination of Maximal Aerobic Speed

Laboratory Session 12.1: VO_{2max}, Velocity at VO_{2max}, and vVO_{2max} Determination (Refer to *Applied Physiology of Exercise* Chapter 5, Balasekaran, Govindaswamy, Lim, Boey, & Ng, 2021)

Prior to the test, the participant will commence warm-up. He/she will jog an easy or slow pace on the treadmill for 3–5 minutes, followed by simple major muscle group stretches. Once the participant is geared up with the necessary equipment for the metabolic measurement, including HR, the researcher can start the VO_{2max} test on the participant (Table 2). For the continuous running and elevation treadmill protocol, start the treadmill speed at 10 $km{\bullet}h^{-1}$ (this start speed can be slower depending on the ability of your participant). Increase the treadmill velocity by 1 $km{\bullet}h^{-1}$ for 6 minutes (to 15 $km{\bullet}h^{-1}$) at every minute. If your participant is still running, increase the treadmill gradient by 2 $\%{\bullet}min^{-1}$ while keeping speed constant until volitional exhaustion of the participant. Encourage your participant to continue running for as long as he possibly can. Determine post-blood lactate immediately after the VO_{2max} test. Or you may use an alternate continuous running treadmill protocol as some participants may not be able to endure the speed with the 2% increase in gradient. Thus, this protocol will start at 10 $km{\bullet}hr^{-1}$ with 1 $km{\bullet}h^{-1}$ increase in speed at every minute till volitional exhaustion. Include a constant 1% gradient as well (Ali, Balasekaran, Hoon, & Gerald, 2017).

Measurement of VO_{2max} and vVO_{2max}: breath by breath, volume of oxygen (VO_2), volume of carbon dioxide (VCO_2), ventilation (VE), respiratory exchange ratio (RER), heart rate (HR), and OMNI rate of perceived exertion (OMNI RPE) will be recorded in the last 15 seconds of each stage (1 minute). VO_{2max} will be determined according to the criteria for attaining VO_{2max} which are as follows (Table 2):

- Plateau in VO_2 (change < 2.1 $mL{\bullet}kg^{-1}{\bullet}min^{-1}$)
- HR ≥ 95% of maximal HR (max HR = 220 – age)
- R ≥ 1.1

For VO_{2max} criteria, refer to American College of Sports Medicine Guidelines, 2018; *Applied Physiology of Exercise* Chapter 5, Balasekaran, Govindaswamy, Lim, Boey, & Ng, 2021.

Table 2. VO_{2max} **determination (continuous running treadmill protocol or continuous running and elevation treadmill protocol, depending on participant analysis).**

Stages	Time	Treadmill Velocity ($km{\cdot}h^{-1}$)	VCO_2 ($L{\cdot}min^{-1}$)	VO_2 ($L{\cdot}min^{-1}$)	VO_2 ($mL{\cdot}kg^{-1}{\cdot}min^{-1}$)	RER	Heart Rate ($beats{\cdot}min^{-1}$)	VE ($L{\cdot}min^{-1}$)	OMNI RPE
1									
2									
3									
4									
5									
6									
7									
8									
9									
10									
11									
12									

Post-blood lactate: ________ $mmol{\cdot}L^{-1}$ (3 marks)

The minimum speed at which maximum oxygen uptake is recorded will be the velocity at maximal oxygen uptake (vVO_{2max}). If the continuous running and elevation treadmill protocol is used to determine VO_{2max}, use the extrapolation method from the submaximal data from Session 2 to derive y = mx + c (function) (Table 5, Figure 5). Use VO_{2max} identified and solve for vVO_{2max}. If a continuous running treadmill protocol is used, you can plot a speed versus VO_2 graph from Table 2 to determine vVO_{2max}.

a) Maximal oxygen uptake (VO_{2max}) test
VO_{2max}: __________ ($mL{\cdot}kg^{-1}{\cdot}min^{-1}$)

(1 mark)

Velocity at VO_{2max} (vVO_{2max}): __________ ($km{\cdot}h^{-1}$) (Refer to Laboratory Session 8 and *Applied Physiology of Exercise* Chapter 5 Figure 6, Balasekaran, Govindaswamy, Lim, Boey, & Ng, 2021)
Interpretations of participant's VO_{2max} test and vVO_{2max}:

(5 marks)

Laboratory Session 12.2: Determination of Lactate Threshold and Velocity at Lactate threshold (Table 3, Refer to Laboratory Session 10) Submaximal till Exhaustion Protocol — Discontinuous Protocol (Gupta & Balasekaran, 2013; Ali, Balasekaran, Hoon, & Gerald, 2017; Balasekaran et al., 2020)

The participant will perform a discontinuous submaximal protocol in order to determine the linear relationship between oxygen uptake at various submaximal exercise intensities (Table 5, Figure 5, Appendix A Figure B).

Prior to the test, the participant will commence warm-up. He/she will jog an easy or slow pace on the treadmill for 3–5 minutes, followed by simple major muscle group stretches. In this session, your participant will perform a series of discontinuous treadmill run lasting 4 minutes for each stage. The treadmill speed will increase at 0.5 $km{\cdot}h^{-1}$ with each stage, which should range between 6–14 km•h^{-1} depending on the ability of your participant and the start or end speed could be higher. Between each stage, your participant will have a 4-minute recovery time. During the recovery stage, the researcher* will take a finger-prick blood sample from the participant to measure blood lactate. Once the blood sample is collected in a small tube (e.g. microvettes, etc.), measure the blood lactate results immediately using the blood lactate machine (portable blood lactate analysers may not indicate accurate results). The blood sample will be collected for blood lactate determination and determination of lactate threshold* (Laboratory Session 10).

Steady state VO_2 for each stage will stabilise around the final 2 minutes of each stage. This will help you develop a linear relationship between VO_2 and treadmill velocity, which will be extrapolated to estimate O_2 cost (energy demand) (Table 5, Figure 5, Appendix A Figure B). In the last minute of each stage, the participant will point to a number on the OMNI RPE scale (Robertson, 2004) and the researcher will record his/her OMNI RPE values. HR values will also be recorded. Exercise will continue till there is a sharp or abrupt nonlinear increase in blood lactate values. When lactate values rise above 4 mmol•L^{-1} for 2 stages consecutively, the test will be terminated (Laboratory Session 10).

After approximately 15 to 20 minutes, recalibrate the metabolic cart and allow the participant to stretch his/her major muscle groups. Once the participant is geared up with the necessary equipment for the metabolic measurement, including HR, the researcher can start the VO_{2max} test on the participant (Table 4). Start the treadmill speed at the same speed as the 9th submaximal treadmill run/10 km•h^{-1} (depending on your participant's start speed). Increase the treadmill velocity by 1 km•h^{-1} for 6 minutes (to 15 km•h^{-1}) (depending on your participant's start speed) at every minute. If your participant is still running, increase the treadmill gradient by 2 %•min^{-1} while keeping speed constant until volitional exhaustion of the participant. Encourage your participant to continue running for a long as he/she possibly can. Determine post-blood lactate immediately after VO_{2max} test.

*Researcher to wear surgical gloves during the test for hygiene purposes. Dispose of laboratory consumables (e.g. finger prick needle, used tissue, etc.) appropriately into the respective disposal bins according to your laboratory's rules and regulations.

The expired air will be sampled through Cosmed Quark b^2 or any metabolic system which allows breath by breath measurement of VE ($L \cdot min^{-1}$), VCO_2 ($mL \cdot min^{-1}$), VO_2 ($mL \cdot min^{-1}$), RER, HR, etc. From the protocols, VO_{2max}, vVO_{2max}, and linear relation between speed and O_2 uptake will be determined (Table 5, Figure 5, Appendix A Figure B).

a) Measurement of VO_{2max} and vVO_{2max}: Breath by breath VO_2, VCO_2, VE, RER, and HR will be recorded and VO_{2max} will be determined according to the criteria for attaining VO_{2max}, which are as follows:
   - Plateau in VO_2 (change < 2.1 $mL \cdot kg^{-1} \cdot min^{-1}$)
   - HR ≥ 95% of maximal HR (max HR = 220 – age)
   - R ≥ 1.1

For VO_{2max} criteria, refer to American College of Sports Medicine Guidelines, 2018; *Applied Physiology of Exercise* Chapter 5, Balasekaran, Govindaswamy, Lim, Boey, & Ng, 2021).

The minimum speed at which maximum oxygen uptake is recorded will be the velocity at maximal oxygen uptake (vVO_{2max}).

This is another measurement of VO_{2max} and vVO_{2max} using submaximal till exhaustion protocol. This is usually done if you do not have time to conduct the VO_{2max} test separately as in Session 1, where a continuous protocol was used to determine VO_{2max} and vVO_{2max}. Continuous and discontinuous protocols have shown similar results (Ali, Balasekaran, Hoon, & Gerald, 2017; Gupta, Balasekaran, & Govindaswamy, 2011).

Question:
Interpretation: Is Session 1 determination of VO_{2max} and vVO_{2max} similar to Session 2's VO_{2max} and vVO_{2max} test? Explain. (5 marks)

Session 2 Results

Fill in Table 3 with your participant's submaximal data (remember to average the data from the final 2 minutes of each workout).

In the table below, record the participant's data from the VO_{2max} test after the 20-minute rest (Table 4).

Use Excel to plot the linear relationship between submaximal treadmill speed and O_2 uptake using a scatter graph with the data from Session 2 of the lab session. For example, a graph is plotted with the following hypothetical data (Figure 5, Table 5).

Now, develop your own graph, similar to Figure 5, using the data obtained during Session 2 and draw the relationship between O_2 uptake and submaximal treadmill speed.

a) Submaximal relationship equation = ____________________

(2 marks)

b) Velocity of VO_{2max} (vVO_{2max})

Using Figure 5, show the extrapolation technique used to obtain vVO_{2max} to determine vVO_{2max}.

vVO_{2max} using the equation = ____________________

(2 marks)

Fill in the following information from the submaximal test conducted during Session 2 (See Appendix A to determine LT. Choose the best method. Refer to Laboratory Session 10 as well).

Table 3. Participant's submaximal data.

Stages	Time	Treadmill Velocity ($km{\cdot}h^{-1}$)	VCO_2 ($L{\cdot}min^{-1}$)	VO_2 ($L{\cdot}min^{-1}$)	VO_2 ($mL{\cdot}kg^{-1}{\cdot}min^{-1}$)	RER	Heart Rate ($beats{\cdot}min^{-1}$)	VE ($L{\cdot}min^{-1}$)	Blood lactate ($mmol{\cdot}L^{-1}$)	OMNI RPE
1										
2										
3										
4										
5										
6										
7										
8										
9										
10										
11										
12										

(3 marks for data)

Table 3. ***(Continued)***

Stages	Time	Treadmill Velocity ($km{\cdot}h^{-1}$)	VCO_2 ($L{\cdot}min^{-1}$)	VO_2 ($L{\cdot}min^{-1}$)	VO_2 ($mL{\cdot}kg^{-1}{\cdot}min^{-1}$)	RER	Heart Rate (beats$\cdot min^{-1}$)	VE ($L{\cdot}min^{-1}$)	Blood lactate ($mmol{\cdot}L^{-1}$)	OMNI RPE
13										
14										
15										
16										
17										
18										
19										
20										
21										
22										
23										
24										

(3 marks for data)

Table 4. Participant's VO_{2max} test data.

Stages	Time	Treadmill Velocity (km•h^{-1})	VCO_2 (L•min^{-1})	VO_2 (L•min^{-1})	VO_2 (mL•kg^{-1}•min^{-1})	RER	Heart Rate (beats•min^{-1})	VE (L•min^{-1})
1								
2								
3								
4								
5								
6								
7								
8								
9								
10								
11								
12								
13								
14								
15								

(3 marks)

Post-blood lactate value: _______ mmol•L^{-1}

Table 5. Plot graph using velocity and maximal oxygen consumption values. Note: These are hypothetical values. Use your own values to plot similar graphs.

Velocity ($km \cdot h^{-1}$)	10	11	12	13	14
VO_2 ($L \cdot min^{-1}$)	2.291	2.559	2.799	3.044	3.451

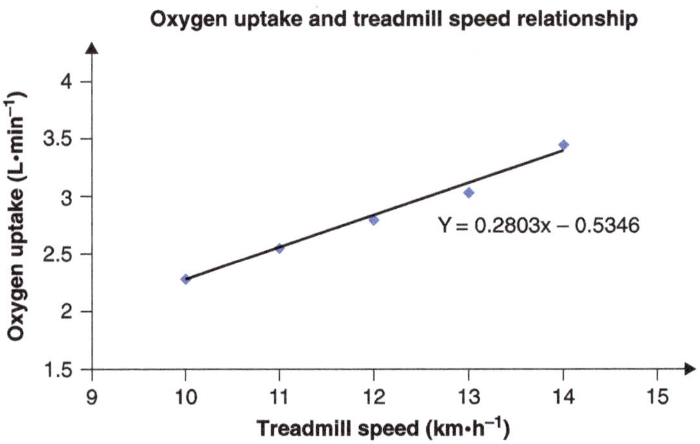

Figure 5. Relationship between O_2 uptake and submaximal treadmill speed.

(6 marks)

Interpretation: VO_{2max} and velocity of VO_{2max} (vVO_{2max}) were determined from the submaximal till exhaustion protocol. Explain and interpret the values.

(6 marks)

2) Velocity at lactate threshold (LT) (vLT): ______ km•h^{-1}

Lactate at LT: ______ mmol•L^{-1}

%VO_{2max} at LT: _______%

HR at LT: _______beats•min^{-1}

%HR_{max} at LT: _______%

(5 marks)

Interpretation of the participant's results from information obtained in (2): (5 marks)

VO_2 Till Exhaustion Test (T_{lim} Test): To Determine MAS (Gupta & Balasekaran, 2013)

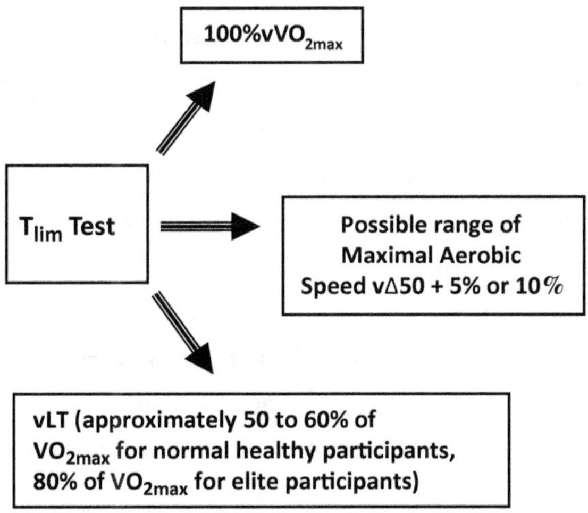

Velocity at Δ50 (vΔ50) is in between 100% vVO_{2max} and velocity at LT (vLT 50 to 60% of VO_{2max}). However, possible range of MAS is vΔ50 + 5% or 10%. Note: Δ50 refers to delta 50.

MAS is in the range of vLT (50 to 60% VO_{2max} for normal healthy participants) and vVO_{2max} (100% of VO_{2max}). vΔ50 is in the middle and MAS is possibly closer to 100% of VO_{2max} as shown in the diagram below. In order to find MAS, find the percentage, which may be vΔ50 + 5% or 10%.

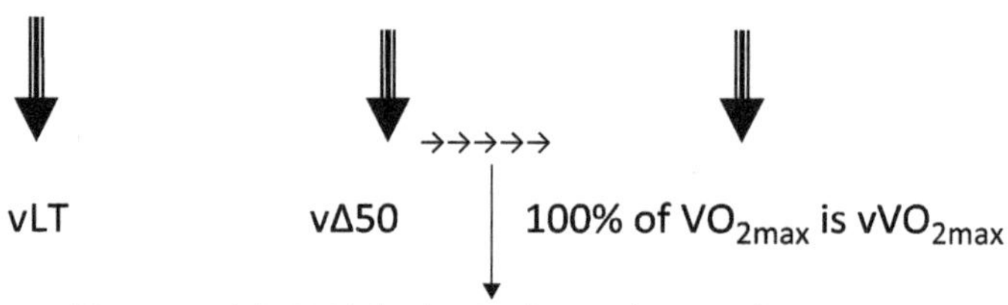

Maximal aerobic speed (MAS) (xx): Find out the % of vVO_{2max}:______

Laboratory Session 12.3: VO_2 Till Exhaustion Test (T_{lim}) at Velocity of 100% VO_{2max} ($Tlim_{vVO2max}$) (Gupta & Balasekaran, 2013) (Table 6)

VO_2 till exhaustion test will be performed at velocity of 100% VO_{2max}. After warm-up, the participant will continue running for as long as possible until they are exhausted in T_{lim} test. VO_2, HR, and RER will be recorded at every 15 seconds in each run. For researchers conducting studies, the blood sample for blood lactate determination will be collected during warm-up, at the 1st, 2nd, 3rd, 4th, and 5th minute during each T_{lim} run. The peak value among the values will be considered as the post-exercise peak blood lactate for all these tests. For teaching laboratory purposes, take post-blood lactate immediately after exercise at every minute for 5 minutes.

Time taken for exhaustion ($Tlim_{vVO2max}$): ____min

(1 mark)

Interpretation of participant's $Tlim_{vVO2max}$:

(5 marks)

Laboratory Session 12.4: VO_2 Till Exhaustion Test vΔ50 + 5% (vΔ50 is Median Velocity of 100% VO_{2max} and Velocity at LT (vLT)) (Gupta & Balasekaran, 2013) (Table 7)

VO_2 till exhaustion test will be performed at vΔ50 + 5%. After warm-up, the participant will continue running for as long as possible until they are exhausted in T_{lim} test. VO_2, HR, and RER will be recorded at every 15 seconds in each run. For researchers conducting studies, the blood sample for blood lactate determination will be collected after warm-up, at the 1st, 2nd, 3rd, 4th, and 5th minute after each T_{lim} run. The peak value among the values will be considered as the post-exercise peak blood lactate for all these tests. For teaching laboratory purposes, take post-blood lactate immediately after exercise every minute for 5 minutes.

Time taken for exhaustion ($Tlim_{Δ50}$ + 5%): _____min

(1 mark)

Interpretation of participant's ($Tlim_{Δ50}$ + 5%):

(5 marks)

Table 6. Results of Laboratory Session 12.3: VO_2 till exhaustion test at velocity of 100% VO_{2max}.

Time (min:s)	Treadmill Velocity (km•h^{-1})	VCO_2 (L•min^{-1})	VO_2 (L•min^{-1})	VO_2 (mL•kg^{-1}•min^{-1})	RER	Heart Rate (beats•min^{-1})	VE (L•min^{-1})	Blood Lactate (mmol•L^{-1})

(3 marks for data)

Lactate values (mmol•L^{-1})
Warm-up: ________
Post-blood lactate: ________
1st min: ________
2nd min: ________
3rd min: ________
4th min: ________
5th min: ________

Table 7. Results of Laboratory Session 12.4: VO_2 till exhaustion test at vΔ50 + 5% (vΔ50 is median velocity of 100% VO_{2max} and velocity at LT (vLT)).

Time (min:s)	Treadmill Velocity (km•h^{-1})	VCO_2 (L•min^{-1})	VO_2 (L•min^{-1})	VO_2 (mL•kg^{-1}•min^{-1})	RER	Heart Rate (beats•min^{-1})	VE (L•min^{-1})	Blood Lactate (mmol•L^{-1})

(*Continued*)

Table 7. (*Continued*)

Time (min:s)	Treadmill Velocity (km•h^{-1})	VCO_2 (L•min^{-1})	VO_2 (L•min^{-1})	VO_2 (mL•kg^{-1}•min^{-1})	RER	Heart Rate (beats•min^{-1})	VE (L•min^{-1})	Blood Lactate (mmol•L^{-1})

(3 mark for data)

Lactate values (mmol•L^{-1})
Warm-up: ________
Post-blood lactate: ________
1st min: ________
2nd min: ________
3rd min: ________
4th min: ________
5th min: ________

Are the T_{lim} test timings for 100% vVO_{2max} and vΔ50 + 5% to exhaustion (time to exhaustion) similar to the literature found? Is the duration too long or too short? Is it appropriate for MAS?

(5 marks)

Test of VO_2 till exhaustion at vVO_{2max} (100%VO_{2max})
vVO_{2max}: _____ km•h^{-1}
$Tlim_{vVO2max}$: _____min
Lactate vVO_{2max}: ______ mmol•L^{-1}

Test of VO_2 till exhaustion at vΔ50 + 5%
vΔ50 + 5%: _____ km•h^{-1}
$Tlim_{v\Delta 50}$ + 5%: ______min
Lactate vΔ50 + 5%: ______ mmol•L^{-1}

(6 marks)

Is vΔ50 +5% an accurate MAS? Explain. Do we need another test of VO_2 till exhaustion at vΔ50 + 10%?

(3 marks)

Laboratory Session 12.5: VO_2 Till Exhaustion Test to Determine vΔ50 + 10% (vΔ50 is Median Velocity of 100% VO_{2max} and Velocity at LT (vLT)) (Gupta & Balasekaran, 2013) (Table 5)

VO_2 till exhaustion test will be performed at vΔ50 + 10%. After warm-up, the participant will continue running for as long as possible until they are exhausted in T_{lim} test. VO_2, HR, and RER will be recorded at every 15 seconds in each run. For researchers conducting studies, the blood sample for blood lactate determination will be collected after warm-up, at the 1st,

Table 8. Results of Laboratory Session 12.5: VO_2 till exhaustion test to determine vΔ50 + 10%.

Time (min:s)	Treadmill Velocity ($km{\cdot}h^{-1}$)	VCO_2 ($L{\cdot}min^{-1}$)	VO_2 ($L{\cdot}min^{-1}$)	VO_2 ($mL{\cdot}kg^{-1}{\cdot}min^{-1}$)	RER	Heart Rate ($beats{\cdot}min^{-1}$)	VE ($L{\cdot}min^{-1}$)	Blood Lactate ($mmol{\cdot}L^{-1}$)

(*Continued*)

Table 8. (*Continued*)

Time (min:s)	Treadmill Velocity (km•h^{-1})	VCO_2 (L•min^{-1})	VO_2 (L•min^{-1})	VO_2 (mL•kg^{-1}•min^{-1})	RER	Heart Rate (beats•min^{-1})	VE (L•min^{-1})	Blood Lactate (mmol•L^{-1})

(3 marks for data)

Lactate values (mmol•L^{-1})

Warm-up: ________

Post-blood lactate: ________

1st min: ________

2nd min: ________

3rd min: ________

4th min: ________

5th min: ________

2nd, 3rd, 4th, and 5th minute after each T_{lim} run. The peak value among the values will be considered as the post-exercise peak blood lactate for all these tests. For teaching laboratory purposes, take post-blood lactate immediately after exercise every minute for 5 minutes.

Below is an example of MAS using *hypothetical* values. $Tlim_{MAS}$ has to be **converted from duration to speed** by utilising your own derived Hill speed-duration relationship curve to determine MAS (Table 9 and Figure 6).

VO_2 till exhaustion test: maximal aerobic speed is $Tlim_{MAS}$ (in order for $Tlim_{MAS}$ to have maximal aerobic contribution, vVO_{2max} has to be subtracted (refer to Gupta & Balasekaran, 2013))

$$Tlim_{MAS} = Tlim_{v\Delta 50*} - Tlim_{vVO2max} = 14 \text{ min } 37 \text{ sec} - 6 \text{ min } 14 \text{ sec}$$

$$Tlim_{MAS} = 8 \text{ min } 23 \text{ sec}$$

Range of $Tlim_{MAS}$: 604.03 ± 125.08 sec

$10 \text{ min} \pm 2 \text{ min } 05 \text{ sec}$

Approximate distance: 3,000 m (refer to Gupta & Balasekaran, 2013)

*Note: $Tlim_{v\Delta 50}$ $\pm 5\%$ or $\pm 10\%$: 747.88 ± 180.72 sec or 12 min 27 sec $\pm$ 3 min (refer to Gupta & Balasekaran, 2013)

Plot a Hill's speed-duration relationship curve using your participant's data. (Please note: do not log data to plot graph, Figure 6, which was derived using hypothetical data from Table 9).

Table 9. Speed data collected.

Speed ($m \cdot s^{-1}$)		Tlim (min)
*MAnS	9.60	0.0625
vVO_2max	4.31	6.23
$v\Delta 50$ + 5% or 10%	3.42	14.62

*Use 1-metre distance using software (e.g. Silicon Coach) to determine maximal anaerobic speed (MAnS) from Laboratory Session 12.6. For example, Table 9, 9.60 $m \cdot s^{-1}$ occurred at 36 m maximal anaerobic speed, therefore Tlim is (36 m/9.60 $m \cdot s^{-1}$)/60 s = 0.0625 min.

You can also use vΔ50 + 5% or 10%, 50 m, 200 m, and 400 m timings and the approximate 33-m peak velocity determined in Laboratory Session 12.6 to get more data points so that the curve will be more accurate. Adding 200 m and 400 m timings might affect the R^2 of the graph as elite athletes are not involved in this protocol and timings can be slow. At times, 200 m and 400 m timings have to be removed to ensure that the R^2 value of the graph is close to 0.99.

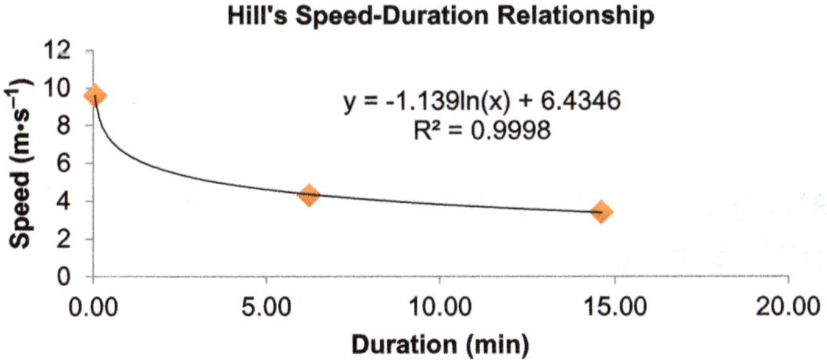

Figure 6. Hill's speed-duration relationship graph (hypothetical data).

(3 marks)

The following equation of the curve: $y = -1.139\ln(x) + 6.4346$ (Figure 6, where x = duration/time to exhaustion (T_{lim}) for the corresponding speed)

Since $Tlim_{MAS}$ = 8 min 23 sec/8.3833 min

Substituting $Tlim_{MAS}$ into the equation: $y = -1.139\ln(x) + 6.4346$,

$MAS = -1.139\ln(8.3833) + 6.4346$

$= 4.0128\ m{\cdot}s^{-1} = 14.45\ km{\cdot}h^{-1}$

$MAS \approx 14.5\ km{\cdot}h^{-1}$

$VO_{2(MAS)}$ (Use your own submaximal regression equation to determine $VO_{2(MAS)}$ (Table 5, Figure 5)) = 0.1743 x 14.5 + 0.6394

$= 3.167\ L{\cdot}min^{-1}$

MAS as $\%VO_{2max}$ (use your own VO_{2max} data determined from Session 1 or 2) = 3.167/3.33075 x 100%

= 95.08%

= 95.08% VO_{2max}

(3 marks)

MAS: __________ $km{\cdot}h^{-1}$

$VO_{2(MAS)}$ = ___________ $L{\cdot}min^{-1}$

MAS as $\%VO_{2max}$ = _________ $\%VO_{2max}$

Note: The percentage of MAS was obtained at 91.08 ± 2.97% vVO_{2max} of total cohort for Gupta & Balasekaran's study (2013).

Interpretations of participant's MAS results:

(3 marks)

Laboratory Session 12.6: Determination of Maximal Anaerobic Speed (Gupta & Balasekaran, 2013)

Maximal velocity test, venue: 400-m track

A 50-m distance will be marked on the track, with a clearly visible start and end line. To measure the maximal velocity, it is necessary to compute speed at very short time intervals (less than 1 second) as maximal speed may occur in a period of less than 1 second. Two timing gates (SMARTtiming, an equipment to measure the time taken to cover a specific distance) will be placed at the start and end line. The timing gates have to also be placed at the 20-m mark, followed by every 10 m (Figure 7). A video camera has to be used to record the 20–50-m part of the sprint. Research has shown that maximal speed occurs between 30–40 m for healthy college participants (Gupta & Balasekaran, 2013). The video recording will be analysed using a software (e.g. Silicon Coach) to determine maximal velocity. A stopwatch will be used to record the overall 50-m timing.

The participant will warm up properly. Following the general warm-up, the participant will also perform 4–5 strides of 40 m with a 3–4 minute recovery time in between strides (mimicking the actual maximal velocity test).

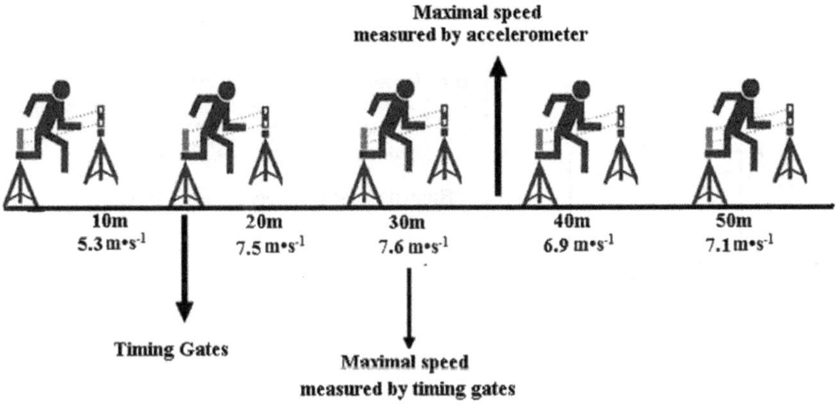

Figure 7. Schematic diagram using timing gates of marking and placing equipment to determine maximal speed during 50-m run test.

Two trials of 50-m sprints will be performed, and a 20-minute rest interval will be given between the 2 trials. The best performance will be recorded to the nearest 0.01 seconds.

The participant will accelerate and cover the whole distance in the fastest possible time. The maximal speed during the run will be recorded.

Rest for 20 minutes.

200-m all out run.

When the participant has recovered, he/she can rest for 45 minutes, following which he/she will run 400 m. Alternatively, if the participant is too tired, another session can be arranged for the 400-m run.

Maximal speed between 30–40 m, 50 m, 200 m, and 400 m can be used as data points for the MAS curve (Figure 6) and the 200 m and 400 m can be used to compare with the Bundle prediction equation of the same event in the later part of the chapter.

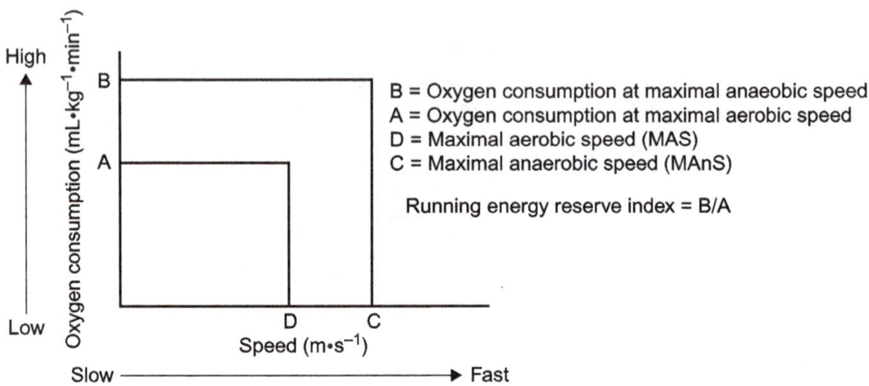

Figure 8. Determination of running energy reserve index.

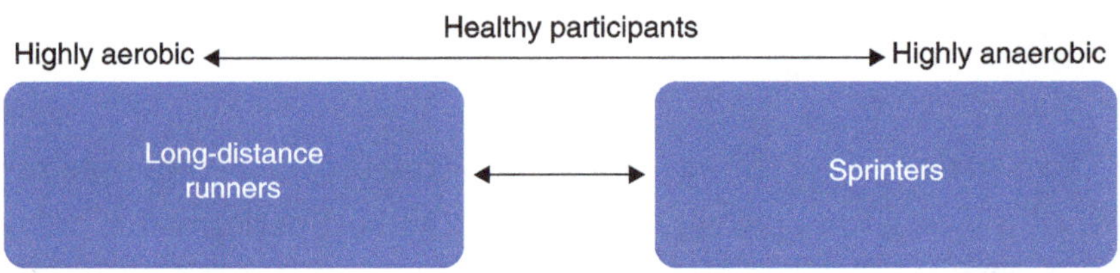

Figure 9. The scale based on RERI among 2 extreme cohorts: sprinters and endurance runners.

Mapping of Running Energy Reserve Index Based on Oxygen Consumption and Maximal Run Speeds (Gupta & Balasekaran, 2013)

After determining MAnS and MAS, their corresponding oxygen consumption will be determined. The index will be computed through the ratio of oxygen consumption at MAnS and MAS as shown in Figures 8 & 10. This will be termed running energy reserve index (RERI). The formula for RERI is as follows:

RERI = maximal anaerobic energy (mL•kg^{-1}•min^{-1})/
maximal aerobic energy (mL•kg^{-1}•min^{-1}) (Figures 8 & 10)

After determining the MAnS and MAS among sprinters and endurance runners, their RERI will be located on the scale (Figure 9). The RERI of these 2 categories of participants will be the 2 extreme points of a scale (Figure 11).

Measurement of RERI with Hypothetical data

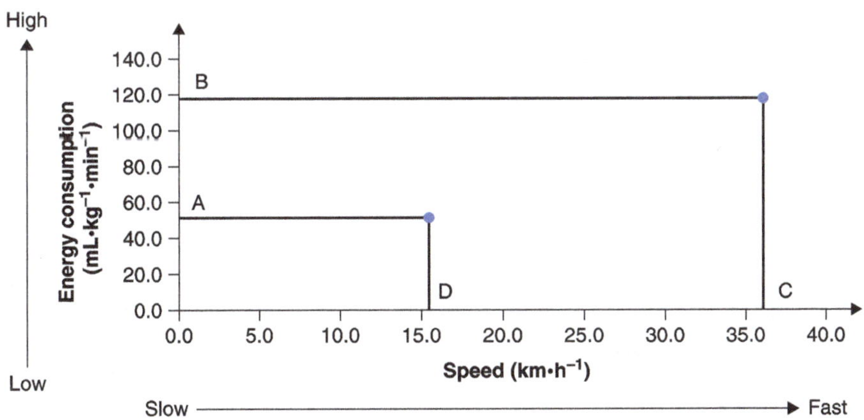

A – Maximal aerobic energy

B – Maximal anaerobic energy

D – Maximal aerobic speed

C – Maximal anaerobic speed

Running energy reserve index (RERI): $\frac{B}{A} = \frac{117.9}{49.3} = 2.39$

Figure 10. Participant's RERI.

Speed Test Results

Maximal anaerobic speed (ManS): 35.8 km$\cdot$h^{-1}

Maximal anaerobic energy: 117.9 mL$\cdot$kg$^{-1}\cdot$min^{-1}

Maximal aerobic speed (MAS): 15.2 km$\cdot$h^{-1}

Maximal aerobic energy: 49.3 mL$\cdot$kg$^{-1}\cdot$min^{-1}

Example of a RERI: 2.39

What is the RERI for your participant? ____________________

(6 marks with diagram)

Comparison of your RERI with world-class runners (Figure 11):

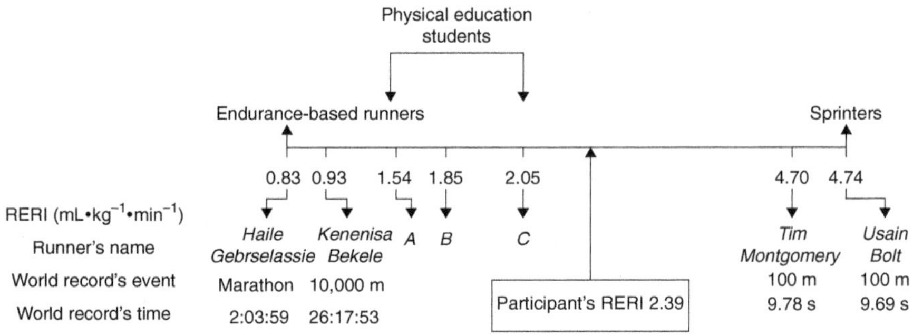

Figure 11. Comparison of world-class runners RERI, as indicated at the extremes. Physical education students' RERI has a broad range of 1.54 to 2.05. The participant's RERI 2.39 is also indicated.

Interpretations of your participant's RERI:

(8 marks)

Bundle's Anaerobic Speed Reserve (AnSR) (Bundle et al., 2003)

vVO_{2max} in km•h^{-1}: ________ (you can use VO_{2max} values from Laboratory Session 12.1 or 12.2, depending on which is more accurate)

(1 mark)

vVO_{2max} is maximal aerobic speed: ______ m•s^{-1}

(1 mark)

Maximal aerobic energy: ______ mL•kg^{-1}•min^{-1} (use the submaximal equation, y = mx + c function, which you used from your data)

(1 mark)

Maximal anaerobic speed during 8-step treadmill: _________ km•h^{-1}
(Refer to Bundle et al., 2003) (1 mark)

Bundle et al. protocol (2003):

Depending on the participant's speed (which can be figured from the maximal speed derived from the 50-m test), start the participant with a modest speed away from his maximal speed. Maximal anaerobic speed will be estimated from the highest speed that the participant will be able to maintain for 8 steps without backward drift on the treadmill. The fastest speed for which 8 steps is completed with no backward drift on the treadmill will be considered the maximal speed by anaerobic power. Give adequate rest in between trials.

Note: watch out for the participant's safety and ensure that the participant puts on a safety harness prior to the start of the test.

Maximal anaerobic speed during 8-step treadmill is maximal anaerobic speed: ________ m•s^{-1} (1 mark)

Maximal anaerobic energy: _______ $mL \cdot kg^{-1} \cdot min^{-1}$ (use your submaximal equation, y = mx + c function, which you used from your data)

(1 mark)

Use Bundle values to calculate RERI: ______________________

(1 mark)

Interpretation: are Bundle and RERI index similar? Explain.

(4 marks)

Can you predict the participant's performance in various sprint events using Bundle's AnSr method? (See example below with hypothetical values)

Bundle's AnSR method requires the following formula (extrapolated from Bundle et al., 2003):

Anaerobic speed reserve = Spdan − Spdaer

= maximum anaerobic speed − maximum aerobic speed (use values previously obtained)

Using hypothetical values to predict performance.

$21.2\ km \cdot h^{-1} - 11.1\ km \cdot h^{-1} = 10.1\ km \cdot h^{-1}$

$= 10.1\ km \cdot h^{-1} \times 1000 \div 60 \div 60$ (change to $m \cdot s^{-1}$)

$= 2.805\ m \cdot s^{-1}$

The following page shows the respective predictions for each of the sprint events using Bundle's AnSR method. The estimated timings for elites may be more accurate than sub-elites or non-elites as Bundle used elite participants in his study.

Sprint Events Estimated Timing

100-m 100 ÷ 2.805 = 35.6502 seconds

200-m 200 ÷ 2.805 = 71.3014 seconds = 1 min 11 sec

400-m 400 ÷ 2.805 = 142.6025 seconds = 2 min 22 sec

AnSR: __________ (refer to Bundle's formula)

(5 marks)

Similar to the example predictions for 100 m, 200 m, and 400 m (300 m can also be used to predict timings), predict the participant's 100 m, 200 m, 400 m, and 800 m timings. Compare the 200 m and 400 m timings which were obtained in Session 6. Are your predictions for the participant's performance similar? Explain.

(6 marks)

Can you predict the participant's performance in various events above 240 seconds using Bundle's AnSr method? Explain your answer.

(3 marks)

Write a summary of the participant's performance with recommendations and conclusions using the laboratory results.

(8 marks)

Applied Question Based on Laboratory Results

How would you translate the knowledge learnt and results of this laboratory for use in training/prediction for healthy adults and children? Give examples of training methods for these two groups based on your understanding of performance prediction.

(12 marks)

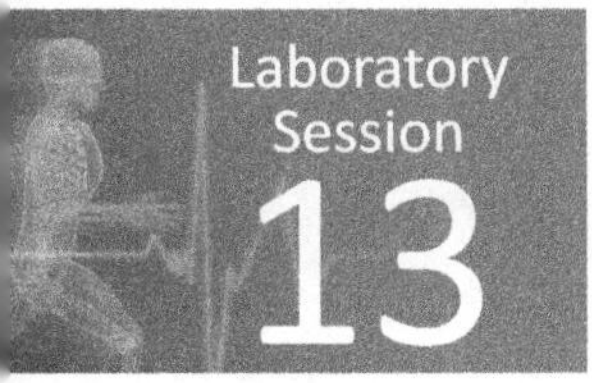

Plasma Volume Changes During Exercise

In this laboratory session, you will learn the procedures to determine the relative change in plasma volume from haematocrit (Hct) responses during exercise. A comparison in Hct responses will be made between two different intensities of the cycle ergometer exercise. You will also determine the relationship between the changes in Hct, blood pressure, and heart rate (HR) during the cycle ergometer exercise. In addition, you will also determine the Hct change during recovery after the exercise.

Cycle Ergometer Exercise with Plasma Volume Measurement

Equipment needed:

1. Cycle ergometer
2. Polar heart rate strap/transmitter
3. Blood lancet
4. Haematocrit tubes
5. Seal for haematocrit tubes
6. Haematocrit centrifuge
7. Stethoscope
8. Sphygmomanometer/blood pressure monitor
9. Electrocardiogram (ECG) (if laboratory has these equipment and students can learn from examining the print outs of the electrical activity of the heart and determine HR during exercise)
10. Electrodes for ECG

Overall Procedure

There will be 2 participants in this laboratory session: one male and one female. Position the participant on the cycle ergometer and adjust his/her seat height. The participant is to be seated and at rest for at least 10 minutes. Record the participant's resting HR and blood pressure and take participant's blood sample as well. During the exercise, record the participant's HR and blood pressure in the last 15 seconds (i.e. 4:45 to 5:00 minutes) of the workload stage (Table 1), followed by taking the participant's blood sample immediately after the exercise. During the recovery stage, record the participant's HR and blood pressure in the last 15 seconds (i.e. 4:45 to 5:00 minutes) and take the participant's last blood sample (Tables 2 & 3).

Table 1. Cycle exercise protocol for males and females.

	Workload/Power			Exercise Period	Recovery Period
	kpm/min	Resistance (kilopond)	Watts		
Males	600	2	98	5 min	5 min
	900	3	147	5 min	5 min
Females	450	1.5	74	5 min	5 min
	600	1.5	98	5 min	5 min

*Pedalling rates must be kept at 50 rpm.

Data Collection Sheet

Name of participant: ______________

Gender: ______________

Height: _________ m

Weight: _________kg

BMI: _____________

Age: __________

Body fat % (optional): __________

Resting HR: __________ beats•min^{-1}

Resting blood pressure: ______________ mmHg

Resting Hct: ____________ %

Table 2. Cycle exercise protocol for males and females.

Workload		Exercise Period	HR (beats · min^{-1})	Blood Pressure (mmHg)	Recovery Period	HR (beats · min^{-1})	Blood Pressure (mmHg)
kpm/min	Watts						
600	98	5 min			5 min		
900	147	5 min			5 min		

Post-exercise Hct: ___________ %

Post-recovery Hct: ___________ %

Table 3. Data collection for female participant.

Workload		Exercise Period	HR (beats · min^{-1})	Blood Pressure (mmHg)	Recovery Period	HR (beats · min^{-1})	Blood Pressure (mmHg)
kpm/min	Watts						
450	74	5 min			5 min		
600	98	5 min			5 min		

Post-exercise Hct: ___________ %

Post-recovery Hct: ___________ %

How to Take Blood?

Place the participant's hand in warm water. Remove the hand from the water and dry it using a towel or clean tissue. Take the blood sample by piercing a fingertip with a blood lancet and filling the haematocrit tube.

Things to Take Note While Taking Blood:

- Allow the blood to flow freely so that squeezing is not necessary
- Wipe away the first drop of blood
- Do not allow the tip of the haematocrit tube (Figure 2) to touch the skin (do not allow it to rest on skin)
- Hold the haematocrit tube at a right angle to sample site
- Draw blood until the graduation mark has been reached
- If the blood is drawn slightly past the mark, touch the tip with a piece of moistened tissue to draw out excess blood
- Insert the haematocrit tube into the seal to close off one end of the tube (Figure 1)

Place Haematocrit Tube in Centrifuge

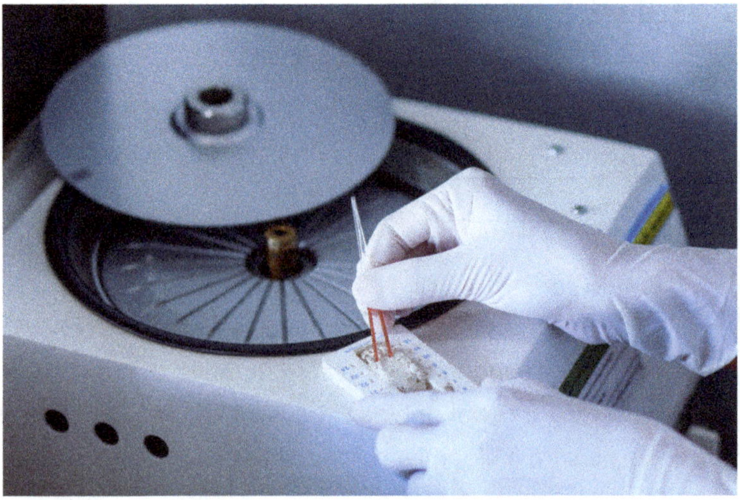

Figure 1. Sample of a haematocrit centrifuge and process of sealing the haematocrit tube.

Figure 2. Haematocrit tubes for blood collection

Centrifuge Procedures (Figure 1)

- Label the position of all the tubes and ensure the centrifuge is balanced
- Place the protective covering over the tubes prior to switching on the centrifuge
- Let the centrifuge tube run for 5 minutes
- Remove the protective cover and cap. Read haematocrit in percentage using a micro-haematocrit reader card usually available in laboratories (Figures 3a & 3b)

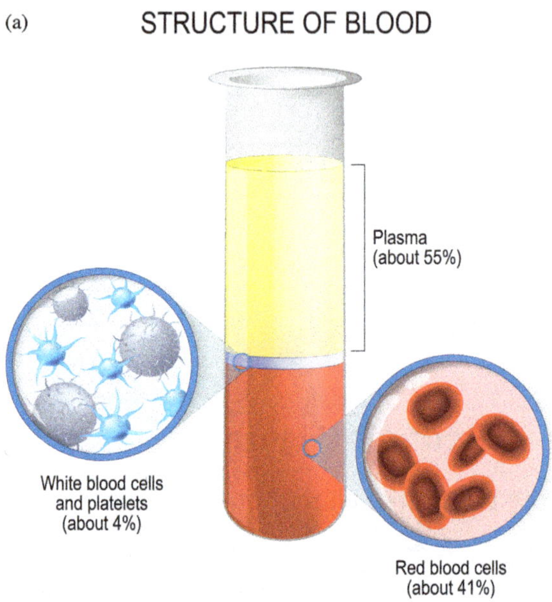

Figure 3a. Blood sample results in a haematocrit tube.

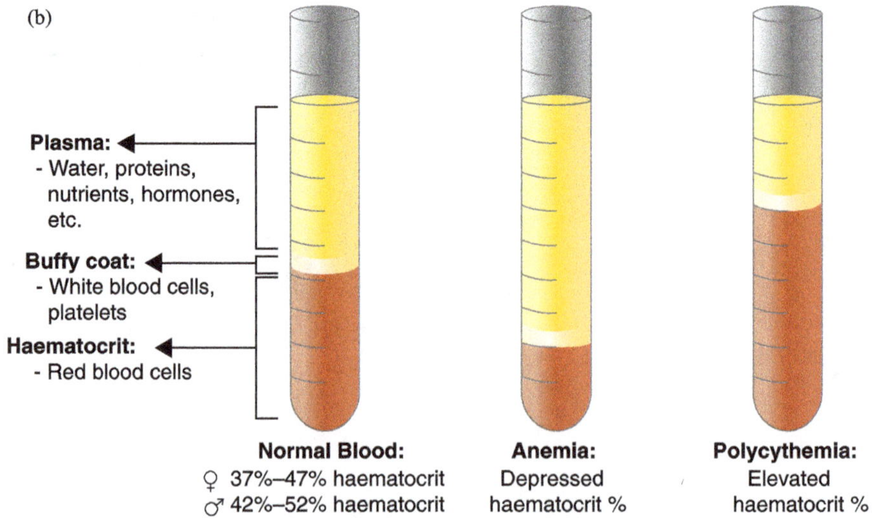

Figure 3b. Samples of haematocrit readings.

Plot the pre-exercise, exercise, and recovery Hct, blood pressure, and HR for both workloads. Calculate the change in plasma volume (PV) that occurred during exercise and recovery (15 marks).

$\% \Delta PV = 100 / 100 - Hct_1 \times 100(Hct_1 - Hct_2) / Hct_2 \%$ (Van Beaumont, 1972)

where:
$\%\Delta PV$: percentage change in plasma volume
Hct_1: haematocrit before exercise
Hct_2: haematocrit after exercise
Procedure: $\%\Delta PV$ calculated is multiplied by the after value to correct it

Questions (40 marks)

1. What information is provided by measuring the Hct during exercise? Is the Hct related to hemoconcentration during physical activity? What are the implications of a changing Hct with regards to oxygen transport functions during exercise? (10 marks)

2. Did the Hct change with the varying exercise intensities? In what direction did it change? Why? (7 marks)

3. Did the Hct change during the recovery period? Why? (8 marks)

4. What are the implications of volume plasma changes for endurance athletes? (5 marks)

5. Are there any gender differences in blood pressure and HR during exercise and recovery? Explain. (5 marks)

6. Are there any gender differences in Hct at rest, after exercise, and recovery? Explain. (5 marks)

Laboratory Session 14

Postprandial Lipaemia

Postprandial Lipaemia

Throughout the course of a day, adults can consume between 30–150 g of triglyceride (TG) along with phospholipids, cholesterol, and fat-soluble vitamins. Nearly all fat ingested during meals is absorbed and enters the blood as TG-rich lipoproteins called chylomicrons (Dubois et al., 1998). Postprandial lipaemia (PPL) relates to the elevation in plasma TG following a fat-containing meal, and is characterised by the long residence time of chylomicron and very-low-density lipoprotein cholesterol (VLDL-C) remnants in circulation (Mero et al., 1998). With regular feedings and the long duration of the hyperlipidaemia, the body is constantly in a postprandial lipemic state. Due to an augmented intestinal blood flow increasing the secretion of pre-formed chylomicrons, simultaneous high-fat meals within 6 hours of each other result in a faster attainment and higher peak TG concentrations (Fielding et al., 1996). The plasma TG response to fat ingestion in the post-absorptive state is related to atherosclerosis. More intense postprandial responses are directly correlated with cardiovascular risk (Romon et al., 1997), with higher postprandial TG levels being found in patients with clinically documented atherosclerosis compared to the control group (Brown et al., 1992).

The Effect of Exercise on Postprandial Lipaemia

Endurance athletes exhibit higher fasting plasma concentrations of high-density lipoprotein cholesterol (HDL-C), and lower plasma TG concentrations than untrained individuals (Grantham & Balasekaran, 2004). The postprandial period tests both trained and untrained individuals' metabolic capacity for TG as the concentrations of TG-rich lipoproteins are at their highest during this time (Hardman & Herd, 1998). PPL, following either an oral fat tolerance test or an intravenous fat tolerance test, is lower in athletes than sedentary men (Cohen, Noakes, & Benade, 1989; Hartung et al., 1993). Endurance-trained males, with high total HDL-C and HDL_2-C levels have an enhanced ability to clear TG during the postprandial period possibly because the mean half-life of chylomicron-TG is shorter in athletes than in sedentary men (Cohen, Noakes, & Benade, 1989). Like oral fat loads, intravenous fat clearance is related to fasting TG concentrations (Sady et al., 1986). In athletes, the lower fasting TG plasma concentrations correspond up to 92% (4.8 $\%{\bullet}min^{-1}$ versus 2.5 $\%{\bullet}min^{-1}$) faster clearance of an intravenously injected (1 $mL{\bullet}kg^{-1}$ body weight) fat emulsion (Sady et al., 1988). Therefore, the athlete's fat clearance was nearly twice that of sedentary men, despite only exhibiting 2% higher lipoprotein lipase (LPL) activity when compared to their heavier, more muscular, sedentary counterparts. However, closer examination of body compositional data suggests that the trained athlete's efficiency to clearance fat could be overestimated two-fold in this study. The athletes were, on average, 17.8 kg lighter than their untrained counterparts. Therefore, in relative terms, the athletes were removing 3.29 compared to 2.15% $fat{\bullet}min^{-1}$, which corresponds to a 53% faster removal rate.

The primary aim of this cross-sectional laboratory protocol is to examine the postprandial lipaemic response in aerobically-trained, strength-trained, and untrained individuals. Aerobic training improves PPL, but little is known about the effect of strength training. It was hypothesised that the lipaemic response would be higher in sedentary/untrained subjects than seen in both aerobic and strength-trained individuals. Support for this hypothesis would add weight to the argument for strength training to be incorporated into a programme to lower the risk of cardiovascular heart disease. The improved lipaemic response of aerobically-trained individuals may be due to differences in body composition and basal metabolic rate (BMR) (Laboratory Session 3), i.e. fat-free mass (FFM) is highly related to BMR and skeletal muscle LPL activity. It was hypothesised that the FFM of strength-trained participants is greater than that of aerobically-trained participants, who have a similar FFM as that of sedentary individuals. Due to the higher FFM in strength-trained individuals, it was also proposed that BMR would be higher in these individuals compared to aerobically-trained participants, who, in turn, would have a greater BMR than sedentary participants. In this laboratory, we will not examine BMR as this is included in Laboratory Session 3. However, researchers can include the BMR assessment in this laboratory to study lipaemic response and its effects by FFM on BMR with individuals trained differently (strength versus aerobic versus untrained).

Experimental Protocol

Equipment needed:

1) Test meals (Table 1)
2) 18-gauge Teflon winged catheter
3) Extension tube (to be attached to catheter)
4) 3M Tegaderm transparent dressing
5) Isotonic saline
6) Plain 3-ml vacutainer
7) 'Nutator' rotary mixer
8) Centrifuge
9) Chemistry analyser
10) Haematocrit tubes
11) Micro-haematocrit reader

Three healthy normolipidaemic volunteers (plasma triglycerides (TG) < 150 $mg{\cdot}dl^{-1}$, plasma concentrations of total cholesterol (TC) < 200 $mg{\cdot}dl^{-1}$, plasma high-density lipoprotein cholesterol (HDL-C) approximately 60 $mg{\cdot}dl^{-1}$), either male or female, can participate in this laboratory session (Note: these range of values may not be satisfied by all participants in a class, but for research purposes, this criteria can be stringent). However, take note that the exact mechanism(s) behind the gender difference in lipolysis during exercise is limited and may be affected by females' menstrual cycle. As such, females need to ensure that they are not within their cycle if they are included as participants. Thus, having solely all male or all female participants would be better to see the differences due to different modes of exercise and/or physiological differences between genders. The 3 volunteers can have different training backgrounds, such as aerobically-trained (AT), strength-trained (ST), and untrained (UT). An untrained individual is classified as a person who exercises 1 or 2 times a week or does not exercise at all.

PPL will be determined following at least a 12-hour fast and 16 hours after the completion of the last exercise bout (any exercise). Participants will have a catheter inserted by a trained phlebotomist into an antecubital vein from which an Interlink System Injection Site (Baxter Healthcare Corp, Illinois, United States) is attached. An 8-ml baseline blood sample will be taken for fasting plasma lipid profile and glucose. Haematocrit (Hct) and haemoglobin concentrations can also collected to determine plasma volume (Dill & Costill, 1974) (optional, discretion by the lecturer but encouraged for researcher; Laboratory Session 13 specifically deals with plasma volume changes during exercise). Plasma volume changes can be used as correction factors to have more accurate results for postprandial analysis (Grantham, Mayo, O'Brien, & Balasekaran, 2004). Over the following 15 minutes (Tsetsonis et al., 1997), participants will consume a test meal (Table 1), composed of approximately 50% carbohydrate (CHO), 42% fat, and 8% protein. A 5-ml blood sample will be drawn 30 minutes postprandially, then hourly for the following 6 hours. All samples will be analysed for lipid profiles, glucose, and plasma volume (optional — discretion by the lecturer but encouraged for researcher as it can be used as correction factor (Grantham, Mayo, O'Brien, & Balasekaran, 2004)). Participants can drink water and engage in low energy expenditure activities such as watching television and reading during the postprandial period. This protocol may also be used to determine the following PPL markers: (1) incremental area under the TG curve; (2) total area under the TG curve; (3) time to peak TG concentration; and (4) return of TG to baseline values (Wideman et al., 1996).

Table 1. Nutritional composition of test meals.

Food	Weight (g)	kcal	Protein (g)	CHO (g)	Fat (g)	Sat. Fat (g)	Chol. (mg)
Pasta Sauce	150	140.63	2.34	21.09	4.68	0.59	—
Pasta	100	131.00	5.15	24.93	1.05	0.15	33
Plain Pringles	50	287.11	1.79	26.95	19.79	5.39	—
Canola Margarine	5	35.71	—	—	3.93	0.36	—
White Bread	100	267.00	8.2	49.5	3.6	0.81	1.00
Magnum Ice Cream	150	334.00	4.9	27.2	22.4	6.10	—
Total	555	1195.45	22.38	149.67	55.45	13.40	1.33

Values are in Weight (g), kcal, Protein (g), Carbohydrate (CHO) (g), Fat (g), Saturated Fat (Sat. Fat) (g), Cholesterol (Chol.) (mg).

Blood Lipids

Before the commencement of all PPL tests, an 18-gauge Teflon winged catheter (Instye®, Becton, Dickinson and Company, Utah, United States) will be inserted into an antecubital vein. After its insertion, a Connecta® 25-cm extension tube (Clinico, Bad Hersfeld, Germany,) will be attached to the catheter. The catheter insertion point and its attachment to the injection site will then be covered with a 3M Tegaderm transparent dressing (3M, Burken, Germany). Following each blood draw, the catheter and injection site will be kept patent with the infusion of isotonic saline (Pharmacia & Upjohn Pty Ltd, Bently, Australia). Before the collection of each postprandial sample, the saline and 2 ml of blood will be drawn into a plain 3-ml Vacutainer (Becton Dickinson and Company, New Jersey, United States) and discarded. For the postprandial tests, two 5-ml blood samples will be drawn 30, 60, 120, 180, 240, 300, and 360 minutes after the consumption of the test meal (use Tables 2, 3, 4, & 5 for data collection). Samples will be drawn into 2 separate Vacutainers (Becton Dickinson and Company, Franklin Lakes, USA) tubes with ethylenediaminetetraacetic acid anti-coagulant. Place the Vacutainers on a 'Nutator' (Becton, Dickinson, and Company, Maryland, United States) rotary mixer for 5 minutes, after which a Vacutainer is used to determine the plasma volume, while the other will be spun for 15 minutes in a Clay Adams® Compact II Centrifuge (Becton, Dickinson and Company, Maryland, United States). Measure plasma TC, TG, HDL-C, and glucose over the 8 time periods of the postprandial session using a dry chemistry analyser (e.g. Kodak Ektachem DT-60II, Eastman Kodak Company, Rochester, USA, or any other newer versions like the Fuji Dry Chem Analyzer).

Plasma Volume

Hct percentages will be analysed twice. Spin the micro-haematocrit capillary tubes for 7 minutes at 3,000 rpm in a centrifuge (Hawksley, England), with haematocrit being determined using a Hawksley Micro-Haematocrit Reader (Hawksley, England). A portable photometric analyser (Optima, HB-202, Japan) will be used to measure haemoglobin concentration. Plasma volume changes will be determined using the Dill and Costill (1974) equation, which is based on the assumption that the volume of circulating blood cells remains constant and that the ratio between the venous haematocrit and whole body haematocrit remains unchanged in all hydrated states. Plasma blood lipid values will be compared with and without adjustment for plasma volume (Laboratory Session 13) (optional — plasma volume changes discretion by lecturer but encouraged for researcher) (use Tables 6 and 7 for data collection).

The Dill & Costill (1974) equation:

$$\text{Plasma volume change} = 100 \times [((\text{Hbpre} \times (100 - \text{Hctpost})) \div ((\text{Hbpost} \times (100 - \text{Hctpre}))] - 100$$

where Hb: haemoglobin

Body Composition

Body composition can be determined using a dual-energy X-ray absorptiometry (DEXA) machine (Hologic QDR-4500W, Hologic Inc, Waltham,

United States) or BIA. The DEXA will allow for the determination of total body weight (BW), body fat percentage (%BF), fat mass (FM), muscle mass (MM), and fat-free mass (FFM). The 7-minute full-body scan will be completed with the participant lying motionless in the anatomical position (Laboratory Session 1).

For Laboratory Session 14, only 3 participants will participate, and students just need to plot the values for all 3 participants with no standard deviation needed (Figure 1). Please note that Figures 1 to 5 refer to 10 participants. The purpose of this laboratory is to highlight the differences in the responses. Researchers can also determine the following PPL markers at their discretion: (1) incremental area under the TG curve (Figure 2); (2) total area under the TG curve (Figure 2); (3) time to peak TG concentration; and (4) return of TG to baseline values (see Figures 2 and 5 below). The total and incremental lipaemic response areas under the curve were calculated using the trapezoidal method (Dubois et al., 1994).

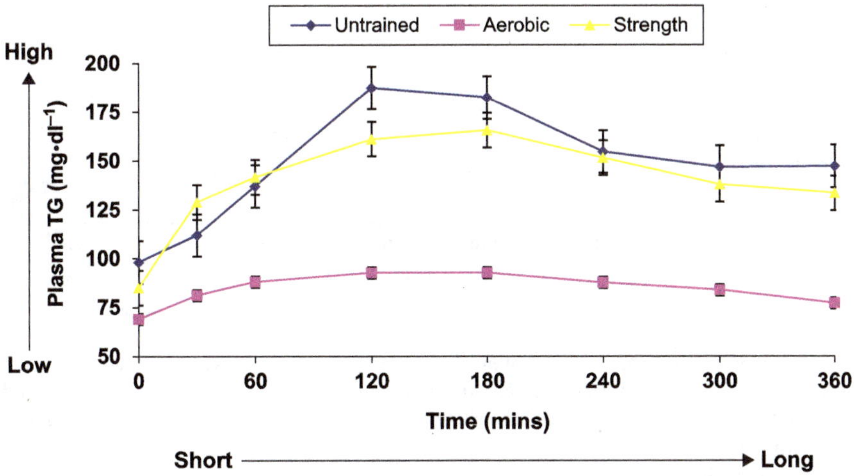

Figure 1. Plasma triglyceride response to fat tolerance tests.

Values are means ± standard deviation for 10 untrained participants (UT), 10 aerobically-trained participants (AT), and 10 strength-trained participants (ST) (Grantham & Balasekaran, 2004).

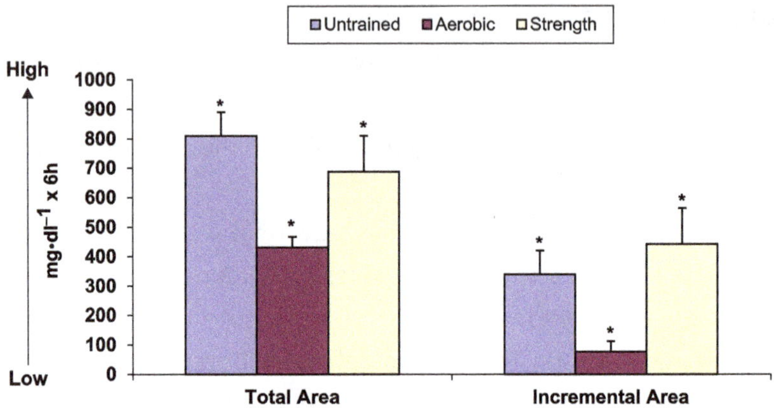

Figure 2. Total and incremental area under the 6-hour plasma TG concentration-time curves. Values are means ± standard deviation for 10 untrained participants (UT), 10 aerobically-trained participants (AT), and 10 strength-trained participants (ST) (Grantham & Balasekaran, 200). *Indicates significant differences $p < 0.05$.

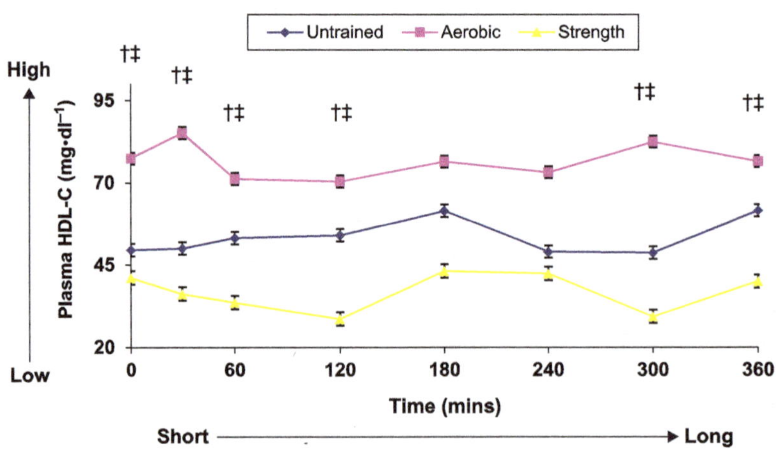

Figure 3. Plasma high-density lipoprotein cholesterol (HDL-C) response to fat tolerance tests. Values are means ± standard deviation for 10 untrained participants (UT), 10 aerobically-trained participants (AT), and 10 strength-trained participants (ST) (Grantham & Balasekaran, 2004).

†AT < ST ($p < 0.05$)

‡AT < UT ($p < 0.05$)

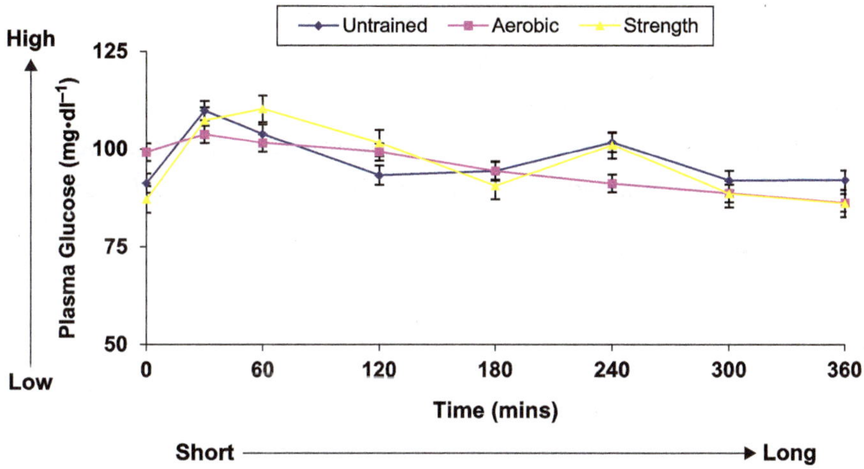

Figure 4. Plasma glucose response to fat tolerance tests.

Values are means ± standard deviation for 10 untrained participants (UT), 10 aerobically-trained participants (AT), and 10 strength-trained participants (ST) (Grantham & Balasekaran, 2004).

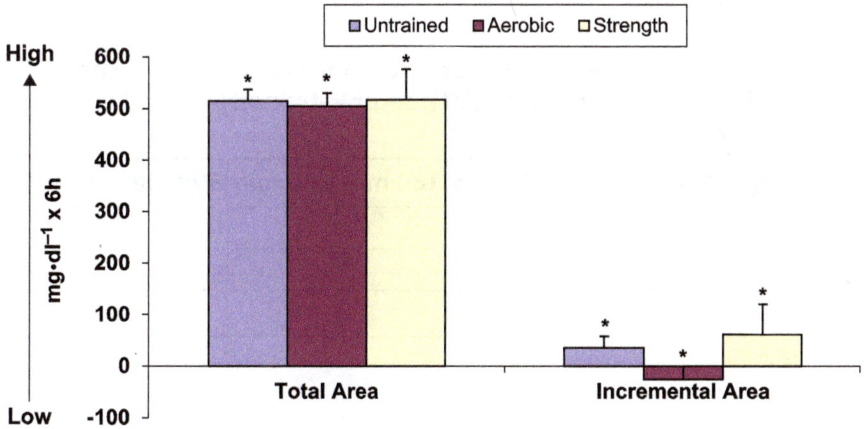

Figure 5. Total and incremental area under the 6-hour plasma glucose concentration-time curves.

Values are means ± standard deviation for 10 untrained participants (UT), 10 aerobically-trained participants (AT), and 10 strength-trained participants (ST) (Grantham & Balasekaran, 2004).

There were significant differences between groups for plasma glucose for incremental area under the curve and total area under the curve.

*Indicates significant differences $p < 0.05$.

There were significant differences between groups for both the plasma triglycerides: incremental area under the curve and total area under the curve with Tukey's post-hoc analysis indicating that incremental area under the curve and total area under the curve were significantly lower in the aerobically-trained than the untrained and strength-trained (both $p < 0.05$) (Figure 2).

Data Collection Sheet

Name of participant: ____________

Gender: ____________

Height: ________ m

Weight: ________kg

BMI: ___________

Age: _________

Body fat %: _________

FFM: _________ kg

BMR/RMR (optional): __________

Table 2. Plasma concentrations of total cholesterol (TC) in fasting and postprandial states for untrained (UT), aerobic-trained (AT), and strength-trained (ST) trials.

	Fasting	30 min	60 min	120 min	180 min	240 min	300 min	360 min
UT								
AT								
ST								

Values are in mg•dl^{-1}.

Table 3. Plasma concentrations of triglycerides (TG) in fasting and postprandial states for untrained (UT), aerobic-trained (AT), and strength-trained (ST) trials.

	Fasting	30 min	60 min	120 min	180 min	240 min	300 min	360 min
UT								
AT								
ST								

Values are in mg•dl^{-1}.

Table 4. Plasma concentrations of high-density lipoprotein cholesterol (HDL-C) in fasting and postprandial states for untrained (UT), aerobic-trained (AT), and strength-trained (ST) trials.

	Fasting	30 min	60 min	120 min	180 min	240 min	300 min	360 min
UT								
AT								
ST								

Values are in mg•dl^{-1}.

Table 5. Plasma concentrations of glucose in fasting and postprandial states for untrained (UT), aerobic-trained (AT), and strength-trained (ST) trials.

	Fasting	30 min	60 min	120 min	180 min	240 min	300 min	360 min
UT								
AT								
ST								

Values are in mg•dl^{-1}.

Table 6. Haemoglobin (Hb) and haematocrit (Hct) concentrations in fasting and postprandial states for untrained (UT), aerobic-trained (AT), and strength-trained (ST) trials.

	Fasting	30 min	60 min	120 min	180 min	240 min	300 min	360 min
UT	Hb Hct							
AT	Hb Hct							
ST	Hb Hct							

Values are haemoglobin (g/dL), haematocrit (percentage, %).

Table 7. Haematocrit (Hct), haemoglobin (Hb) values, and plasma volume changes (Δ) in fasting and postprandial states for during pre- and post-exercise in untrained (UT), aerobic-trained (AT), and strength-trained (ST) participants.

	Time (min)	Hct (%)	Hb (g/dL)	Δ Plasma Volume
UT Pre-exercise	0			
	30			
	60			
	120			
	180			
	240			
	300			
	360			
UT Post-exercise	0			
	30			
	60			
	120			
	180			
	240			
	300			
	360			
AT Pre-exercise	0			
	30			
	60			
	120			
	180			
	240			
	300			
	360			
AT Post-exercise	0			
	30			
	60			
	120			
	180			
	240			
	300			
	360			

(*Continued*)

Table 7. (*Continued*)

	Time (min)	Hct (%)	Hb (g/dL)	Δ Plasma Volume
ST Pre-exercise	0 30 60 120 180 240 300 360			
ST Post-exercise	0 30 60 120 180 240 300 360			

Questions (40 marks)

1. Plot the fasting plasma TG, plasma glucose concentrations, and high-density lipoprotein cholesterol (HDL-C) for the 3 participants, similar to the graphs in Figures 1, 3, and 4. (Figures 2 and 5 are optional — lecturer's discretion) (5 marks)
2. Did the aerobic-trained and strength-trained participants exhibit lower fasting plasma TG, plasma glucose concentration, and HDL-C responses compared to untrained participants? Explain. (10 marks)
3. Between the aerobic-trained and strength-trained participants, which participant exhibited lower fasting plasma TG plasma, glucose concentration, and HDL-C responses? Explain. (10 marks)
4. Which exercise programme is appropriate to reduce postprandial lipemic responses? Explain your answer with physiological rationale. (5 marks)
5. For children in schools, what kind of energy system training should we advocate for the reduction in postprandial lipemic response? Give some examples of the training appropriate for children. (10 marks)

Optional (30 marks)

6. Are there any plasma volume changes between the 3 participants? Explain. (10 marks)
7. Are there any differences between the 3 participants after plasma volume corrections? Explain. (10 marks)
8. Are there any differences between the 3 participants for Hct and Hb? Explain. (10 marks)

Laboratory Session 15

Running Economy Laboratory

There are instances where participants had similar maximal oxygen consumption (VO_{2max}) values but markedly different energy needs during distance running (Table 1).

- World-class male distance runner Alberto Salazar recorded a VO_{2max} of 78 $mL{\cdot}kg^{-1}{\cdot}min^{-1}$, nearly identical to multiple Boston marathon winner Bill Rodgers.
- In 1980, world-class female distance runner, Grete Waitz from Oslo, Norway recorded one of the highest VO_{2max} values (73.5 $mL{\cdot}kg^{-1}{\cdot}min^{-1}$) observed for a woman.
- Derek Clayton, a 2 hour, 8 minutes, and 33 seconds (2.08.33) marathon runner, recorded a 69 $mL{\cdot}kg^{-1}{\cdot}min^{-1}$ VO_{2max} during a laboratory treadmill testing in 1970. Most marathon times hovered around 2.09.00 hours, and it took a while to break Clayton's record. Rob de Castella broke the record with a time of 2.08.18 in 1981 after 12 years. Following which, Alberta Salazar had a time of 2.08.13. When Salazar ran this timing, which was faster than Clayton's, he had a VO_{2max} of 78 $mL{\cdot}kg^{-1}{\cdot}min^{-1}$. It was at least 10 $mL{\cdot}kg^{-1}{\cdot}min^{-1}$ more than Derek Clayton's VO_{2max}. Clayton's success had been attributed to his running economy, as he specialised in the shuffle running technique where his foot rarely came off the road, thus saving precious time as opposed to the high foot lifts. **"Through miles and miles of training, I honed my leg action to such a degree that I barely lifted my leg off the ground."** – Derek Clayton.

- Example of three athletes with similar VO_{2max} values but with greatly different standard-marathon timings:

Alberto Salazar (United States) — VO_{2max}: 78 $mL{\cdot}kg^{-1}{\cdot}min^{-1}$, marathon timing: 2.08.13
Grete Waitz (Norway) — VO_{2max}: 73.5 $mL{\cdot}kg^{-1}{\cdot}min^{-1}$, marathon timing: 2.25.29
Cavin Woodward (United Kingdom) — VO_{2max}: 74.2 $mL{\cdot}kg^{-1}{\cdot}min^{-1}$, marathon timing: 2.19.50

Why do these athletes have markedly different marathons times but have similar VO_{2max} values? Could it be explained by a term called running economy? The graph depicting the running speed and oxygen cost of consumption of 5 great distance runners (Figure 1) below may explain running economy. From Figure 1, are you able to identify who has the best running economy and who has the least?

The estimation of caloric intake and expenditure may explain their energy cost during running. We can calculate these values, compare the athletes, and determine who has a lower energy cost. A lower energy cost may lead to better running timings as it is related to better running economy. Two runners can be compared with such calculations to determine who has a lower energy cost (examples on the next page).

Definition

- **Caloric equivalent**: the number of kilocalories produced per liter of oxygen consumed
- **Caloric cost**: energy expenditure of an activity performed for a specific period of time
- **Caloric cost ($kcal{\cdot}min^{-1}$)**: O_2 consumed ($L{\cdot}min^{-1}$) × caloric equivalent ($kcal{\cdot}LO_2$)

Example:

RER = 0.91

From the treadmill test, O_2 consumed = 2.15 L•min^{-1}

Calculate caloric cost for 30 minutes of exercise.

The caloric equivalent for a RER of 0.91 is 4.936 kcal•LO_2 (see Laboratory Session 3 Table 8)

2.15 O_2•min^{-1} × 4.936 kcal•LO_2 = 10.61 kcal•min^{-1}

10.61 × 30 min = 318.3 kcal

Fractional utilisation of aerobic capacity may also explain superior running economy (Laboratory Session 7; *Applied Physiology of Exercise* Chapter 5, Balasekaran, Govindaswamy, Lim, Boey, & Ng, 2021).

- If 2 runners (A and B) have different or similar aerobic fitness, their fractional utilisation of VO_{2max} may differ. For example, if they have a VO_{2max} of 70 versus 65 mL•kg^{-1}•min^{-1}, respectively, due to the difference in their aerobic capacities, the demands placed on their cardiovascular and muscular systems would be markedly different.
- Runner A, for example, would be working at 83% of his VO_{2max}, whereas runner B would only use 71% of his aerobic capacity.
- Runner B could sustain that pace for a longer period of time and feel less distressed compared to runner A, and may win an endurance race.
- However, this could be interpreted differently as running closer to percentage of VO_{2max} may mean running at a faster speed. Thus, some runners can sustain this fast pace with ease and win a race as the running at a lower percentage to VO_{2max} may mean running at a slower speed.
- Running at closer to percentage of VO_{2max} may also mean affecting running economy as seen in overtrained athletes (Figure 2). Extremely high volume and high intensity may lead to such scenarios and effect running economy with no change in VO_{2max} (Balasekaran, 1993, 2001).

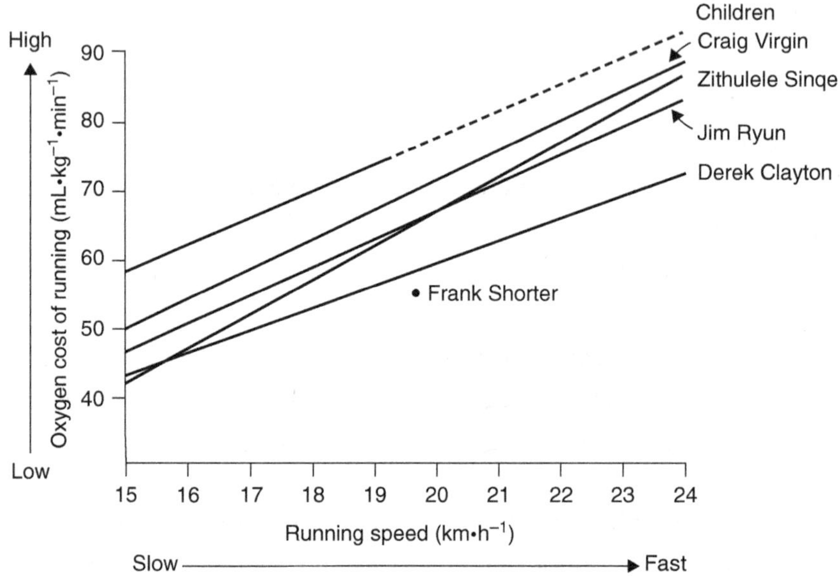

Figure 1. Linear regression line graphs of running speed versus oxygen cost of running across different athletes and children (adapted from Noakes, 2003).

- Figure 1 shows that the men depicted had similar VO_{2max} values but markedly different energy needs during distance running. A large part of the advantage in competition can be attributed to greater running efficiency or running economy. (Note: The data for children in this graph has been extrapolated. Do note their running economy as it shows that they have not reached a high standard of running efficiency. Please observe this while coaching them.)

Running performance can be related with fast timings in 10 km or other distances if one's running efficiency or running economy is superior (see *Applied Physiology of Exercise* Chapter 5, Balasekaran, Govindaswamy, Lim, Boey, & Ng, 2021). Additionally, VO_{2max} is necessary for world-class performances. However, running economy can play a role, as seen by Clayton's fast marathon timings and one of the lowest

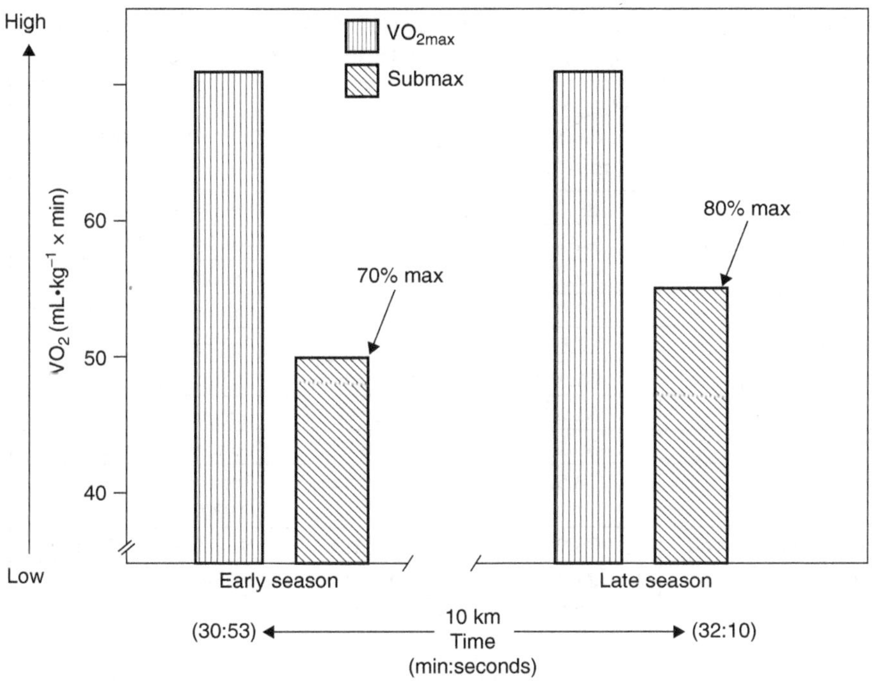

Figure 2. Oxygen uptake during maximal and submaximal running in two periods of seasons (adapted from Dill & Costill, 1974).

VO_{2max} among other top marathon elite runners (69 mL•kg^{-1}•min^{-1}). As mentioned earlier, the highest VO_{2max} for women is Grete Waitz with 73 mL•kg^{-1}•min^{-1}, which is highly commendable as she is comparable to some of the elite males. In general, women have physiological differences such as lower haemoglobin, smaller heart size, lower lean mass, and higher body fat percentage as compared to males (*Applied Physiology of Exercise* Chapter 5, Balasekaran, Govindaswamy, Lim, Boey, & Ng, 2021). Table 1 indicates the VO_{2max} values in other elite endurance athletes, accompanied by their major performances.

In this lab, you will have two participants. Both participants will run a discontinuous submaximal VO_2 test followed by a 20-minute rest to determine VO_{2max} test (same as Laboratory Session 7).

Table 1. Maximal oxygen consumption (VO_{2max}) in elite endurance athletes and their major performances (adapted from Noakes, 2003).

Athlete	VO_{2max} Value ($mL \cdot kg^{-1} \cdot min^{-1}$)	Major Performance	Reference
Dave Bedford	85.0	10 km WR 1973	Berg (1982)
Steven Prefontaine	84.4	1 mile 3:54.6	Pollock (1977)
Gary Tuttle	82.7	2:17 marathon	Pollock (1977)
Kip Keino	82.0	2 km WR 1965	Saltin & Astrand (1967)
Don Lash	81.5	2 mile WR 1937	Robinson et al. (1937)
Craig Virgin	81.1	2:10:26 marathon	Curreton et al. (1975)
Jim Ryun	81.0	1 mile WR 1967	Daniels (1974)
Steve Scott	80.1	1 mile 3:37.69	Conley et al. (1984)
Bill Rodgers	78.5	2:09:27 marathon	Rodgers & Concannon (1982)
Matthews Temane	78.0	21.1 km WR 1987	Noakes et al. (1990b)
Don Kardong	77.4	2:11:15 marathon	Pollock (1977)
Tom O'Reilly	77.0	927 km in 6-day race	Davies & Tompson (1979)
John Landy	76.6	1 mile WR 1954	Astrand (1955)
Alberto Salazar	76.0	Marathon WR 1981[ψ]	Costill (1982)
Johnny Halberstadt	74.4*	2:11:44 marathon	Wyndham et al. (1969)
Amby Burfoot	74.3	2:14:28 marathon	Costill & Winrow (1970)
Cavin Woodward	74.2	48–160 km WR 1975	Davies & Tompson (1979)
Kenny Moore	74.2	2:11:36 marathon	Pollock (1977)
Bruce Fordyce	73.3*	80 km WR 1983	Jooste et al. (1980)
Grete Waitz	73.0	Marathon WR 1980	Costill & Higdon (1981)
Buddy Edelen	73.0	Marathon WR 1963	Dill et al. (1967)
Peter Snell	72.3	1 mile WR 1964	Carter et al. (1967)
Zithulele Sinqe	72.0	2:08:05 marathon	Noakes et al. (1990b)
Frank Shorter	71.3	2:10:30 marathon	Pollock (1977)
Willie Mtolo	70.3	2:08:15 marathon	Noakes et al. (1990b)
Derek Clayton	69.7	Marathon WR 1969	Costill et al. (1971b)

WR = World record

*Predicted sea-level values from measurements recorded at medium altitude (5,784 feet) by adding 11%.

[ψ]Subsequently not ratified (short course).

Laboratory Session 15.1: Submaximal Efficiency Tests and VO_{2max} Determination (Gupta & Balasekaran, 2013; Ali, Balasekaran, Hoon, & Gerald, 2017; Balasekaran et al., 2020)

Equipment needed:

1. Treadmill: set at a gradient of 1% to reflect the energy cost of running outdoors
2. Metabolic cart (to measure oxygen consumption; please ensure that it is calibrated)
3. Polar heart rate monitor transmitter
4. Spirometry head gear
5. Nose clip
6. OMNI RPE scale (Robertson, 2004)
7. Safety harness
8. Water for participant

Prior to the test, the participant will commence warm-up. He/she will jog an easy or slow pace on the treadmill for 3–5 minutes, followed by simple major muscle group stretches. This session will involve your participant performing a series of discontinuous treadmill runs lasting 4 minutes for each stage. The treadmill speed will increase at 0.5 $km{\cdot}h^{-1}$ with each stage, which should range between 6–14 $km{\cdot}h^{-1}$ depending on the ability of your participant and the start or end speed could be higher. Between each stage, your participant will have a 4-minute recovery time. During the recovery stage, the researcher* will take a finger-prick blood sample from the participant to measure blood lactate. Once the blood sample is collected in a small tube (e.g. microvettes, etc.), measure the blood lactate results immediately using the blood lactate machine (portable blood lactate analyser may not indicate accurate results).

*Researcher to wear surgical gloves during test for hygiene purposes. Dispose of laboratory consumables (e.g. finger prick needle, used tissue, etc.) appropriately into the respective disposal bins according to your laboratory's rules and regulations.

Steady state VO_2 for each stage will stabilise around the final 2 minutes of each stage. This will help you develop a linear relationship between VO_2 and treadmill velocity (Laboratory Session 7 Figure 5, Appendix A). In the last minute of each stage, the participant will point to a number on the OMNI RPE scale (Robertson, 2004) and the researcher will record his/her OMNI RPE values. HR values will also be recorded. Exercise will continue till there is a sharp increase in blood lactate values, following which the test will terminate. The test will terminate when the desired lactate has been determined. Desired lactate can refer to the following:

1. When plotting a speed versus lactate graph, the linear graph shows a breakaway point or abrupt non-linear rise (Refer to Appendix A)
2. When lactate values rise above 4 $mmol{\cdot}L^{-1}$ for 2 consecutive stages

Laboratory Session 15.2

After approximately 20 minutes, recalibrate the metabolic cart and allow your participant to stretch his/her major muscle groups again. Once the participant is geared up with the necessary equipment for metabolic measurement, including HR, start the VO_{2max} test on your participant. Start the treadmill speed at the same speed as the 9th submaximal treadmill run/10 $km{\cdot}h^{-1}$ (depending on your participant's start speed). Increase the treadmill velocity by 1 $km{\cdot}h^{-1}$ for 6 minutes (to 15 $km{\cdot}h^{-1}$) (depending on your participant's start speed) at every minute. If your participant is still running, increase the treadmill gradient by 2 $\%{\cdot}min^{-1}$ with constant speed until volitional exhaustion of the participant (HR ≥ 95% of maximal HR (max HR = 220 – age)) (For VO_{2max} criteria, refer to American College of Sports Medicine Guidelines, 2018; *Applied Physiology of Exercise* Chapter

5, Balasekaran, Govindaswamy, Lim, Boey, & Ng, 2021). Encourage your participant to continue running for a long as he/she possibly can.

Fill in the following table with your participant's submaximal data (remember to average the data from the final 2 minutes of each work bout).

The data sheet for Participants 1 and 2 are provided below (Tables 2, 3, & 4).

Name of Participant 1: ______________________________

Weight: ________ Height: __________ BMI: __________ Age: ________

Resting HR: _________ Predicted Max HR: __________ 95% of Maximal HR: ___________

Table 2. Participant's submaximal data.

Stages	Time	Treadmill Speed ($km \cdot h^{-1}$)	VCO_2 ($L \cdot min^{-1}$)	VO_2 ($L \cdot min^{-1}$)	VO_2 ($mL \cdot kg^{-1} \cdot min^{-1}$)	RER	VE ($L \cdot min^{-1}$)	Heart Rate ($beats \cdot min^{-1}$)	OMNI RPE	Blood Lactate ($mmol \cdot L^{-1}$)
1										
2										
3										
4										
5										
6										
7										
8										
9										
10										
11										

(*Continued*)

Table 2. (*Continued*)

Stages	Time	Treadmill Speed ($km \cdot h^{-1}$)	VCO_2 ($L \cdot min^{-1}$)	VO_2 ($L \cdot min^{-1}$)	VO_2 ($mL \cdot kg^{-1} \cdot min^{-1}$)	RER	VE ($L \cdot min^{-1}$)	Heart Rate ($beats \cdot min^{-1}$)	OMNI RPE	Blood Lactate ($mmol \cdot L^{-1}$)
12										
13										
14										
15										
16										
17										
18										
19										
20										
21										
22										
23										
24										
25										
26										
27										
28										
29										
30										

Laboratory Session 15.2 Results

In the table below, record the participant's data from the VO_2 maximal test.

Table 3. Participant's data from VO_{2max} test.

Stages	Time	Treadmill Speed (km·h^{-1})	VCO_2 (L·min^{-1})	VO_2 (L·min^{-1})	VO_2 (mL·kg^{-1}·min^{-1})	RER	VE (L·min^{-1})	Heart Rate (beats·min^{-1})

Post-blood lactate value: _______ mmol·L^{-1}

Fill in the following table with your participant's submaximal data (remember to average the data from the final 2 minutes of each work bout).

Name of Participant 2: ______________________

Weight: ________ Height: __________ BMI: __________ Age: ________

Resting HR: _________ Predicted Max HR: __________ 95% of Maximal HR: ___________

Table 4. Participant's submaximal data.

Stages	Time	Treadmill Speed ($km{\cdot}h^{-1}$)	VCO_2 ($L{\cdot}min^{-1}$)	VO_2 ($L{\cdot}min^{-1}$)	VO_2 ($mL{\cdot}kg{\cdot}^{-1}{\cdot}min^{-1}$)	RER	VE ($L{\cdot}min^{-1}$)	Heart Rate ($beats{\cdot}min^{-1}$)	OMNI RPE	Blood Lactate ($mmol{\cdot}L^{-1}$)
1										
2										
3										
4										
5										
6										
7										
8										
9										
10										

(*Continued*)

Table 4. (*Continued*)

Stages	Time	Treadmill Speed ($km{\cdot}h^{-1}$)	VCO_2 ($L{\cdot}min^{-1}$)	VO_2 ($L{\cdot}min^{-1}$)	VO_2 ($mL{\cdot}kg{\cdot}^{-1}{\cdot}min^{-1}$)	RER	VE ($L{\cdot}min^{-1}$)	Heart Rate ($beats{\cdot}min^{-1}$)	OMNI RPE	Blood Lactate ($mmol{\cdot}L^{-1}$)
11										
12										
13										
14										
15										
16										
17										
18										
19										
20										
21										
22										
23										
24										
25										
26										
27										
28										
29										
30										

Laboratory Session 15.2 Results

In the table below, record the participant's data from the VO_2 maximal test.

Table 5. Participant's data from VO_{2max} test.

Stages	Time	Treadmill Speed ($km{\cdot}h^{-1}$)	VCO_2 ($L{\cdot}min^{-1}$)	VO_2 ($L{\cdot}min^{-1}$)	VO_2 ($mL{\cdot}kg^{-1}{\cdot}min^{-1}$)	RER	VE ($L{\cdot}min^{-1}$)	Heart Rate ($beats{\cdot}min^{-1}$)

Post-blood lactate value: ________ $mmol{\cdot}L^{-1}$

Questions (40 marks)

1. What was the VO_{2max} of both participants? Who had a higher VO_{2max} and why? (3 marks)
2. Draw both participants' submaximal data on a graph with VO_2 ($L{\bullet}min^{-1}$) against speed. Who has a better running economy? Explain why they have different running efficiencies based on your graph (refer to Appendix A Figure B). (7 marks)
3. Choose a RER at the same speed and calculate the caloric cost for 30 minutes of exercise. Who has a lesser energy cost and why? (Hint: Refer to example on page 213) (6 marks)
4. Select a speed and calculate the percentage of VO_{2max} for the 2 participants. Who is more efficient and who will win an endurance race? Explain why. (7 marks)
5. What ways can you improve the running economy of an individual? (3 marks)
6. Can running economy be used for other sports? How will you know if one has a good running economy in another sport other than running? Is there such a thing as swimming economy or basketball economy? Explain your answer. (5 marks)
7. What can you do to improve a child's running economy? Explain with examples. (3 marks)
8. Refer to data below on Runner A and B to answer this question. Outline cardiovascular, metabolic, and other factors which underlie the differences between Runner A and B (use your knowledge of other physiological reasons from previous laboratories; refer to *Applied Physiology of Exercise* Chapter 7, Balasekaran, Govindaswamy, Lim, Boey, & Ng, 2021) (6 marks)

Runner A	**Runner B**
VO_{2max}: 75 $mL{\bullet}kg^{-1}{\bullet}min^{-1}$	VO_{2max}: 60 $mL{\bullet}kg^{-1}{\bullet}min^{-1}$
Data below taken at 14 $km{\bullet}hr^{-1}$	
• VO_2 = 48 $mL{\bullet}kg^{-1}{\bullet}min^{-1}$	VO_2 = 55 $mL{\bullet}kg^{-1}{\bullet}min^{-1}$
• Heart rate = 145 $beats{\bullet}mim^{-1}$	Heart rate = 182 $beats{\bullet}min^{-1}$
• RER = 0.87	RER = 1.00
• Blood lactate = 2 $mmol{\bullet}L^{-1}$	Blood lactate = 6 $mmol{\bullet}L^{-1}$
• $VE.VO_{2max}$ = 30 $L{\bullet}min^{-1}$	$VE.VO_{2max}$ = 45 $L{\bullet}min^{-1}$

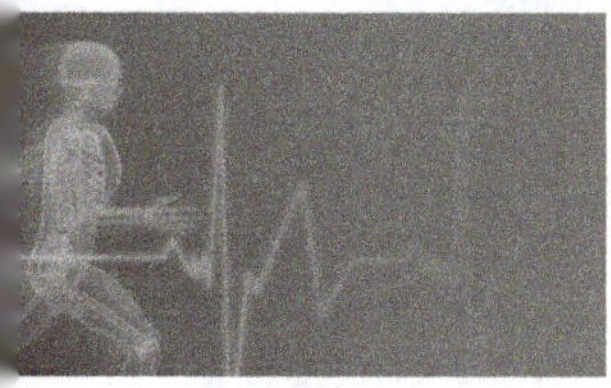

Appendix A (Laboratory Session 12)

1. Log-log Plot Method Using Sample Data to Detect V_{pt} (Thor & Balasekaran, 2012)

1.1. Average the last 2 minutes of the participant's VO_2 ($L{\cdot}min^{-1}$) data for each stage (Figure A).

Time	VO_2
	STPD
min	L/min

0.19283333	0.47279984
0.36633331	0.48771083
0.50916666	0.29653463
0.685	0.47676203
0.85983336	0.47699058
1.0023334	0.53378314
1.17583346	0.65358061
1.34916687	0.68600214
1.50150037	0.793733
1.67666698	1.12755179
1.84533358	1.17279267
2.01116681	1.76634753
2.20850015	1.51295543
2.37066674	1.38840091
2.50600004	1.75853968
2.68883348	2.3149364
2.86583328	1.4966861
3.00883317	1.99106467
3.17266631	2.28290367
3.33849955	2.34517336
3.50799942	1.97152853
3.67383242	2.11618996
3.84199882	1.94572949
4.00516558	2.09196544
4.19699955	1.97852397
4.35999918	1.55263186
4.5236659	2.21018529
4.67366552	2.34894538
4.83933163	1.79739487
	2.05268637

Figure A. Sample of participant's first-stage data. Average the last 2 minutes of the participant's VO_2 values ($L{\cdot}min^{-1}$).

Put the values in a table form (Table A.1).

Table A.1. Sample hypothetical data of participant's speed (km•h^{-1}) versus VO_2 (L•min^{-1}).

Speed (km•h^{-1})	6	6.5	7	7.5	8	8.5	9	9.5	10
VO_2 (L•min^{-1})	2.05	2.12	2.31	2.52	2.76	2.98	3.16	3.23	3.44

1.2. Plot a linear regression equation graph for the participant's speed (km•h^{-1}) versus VO_2 (L•min^{-1}) (Figure B). Ensure that the R^2 value is as close to 0.99 as possible.

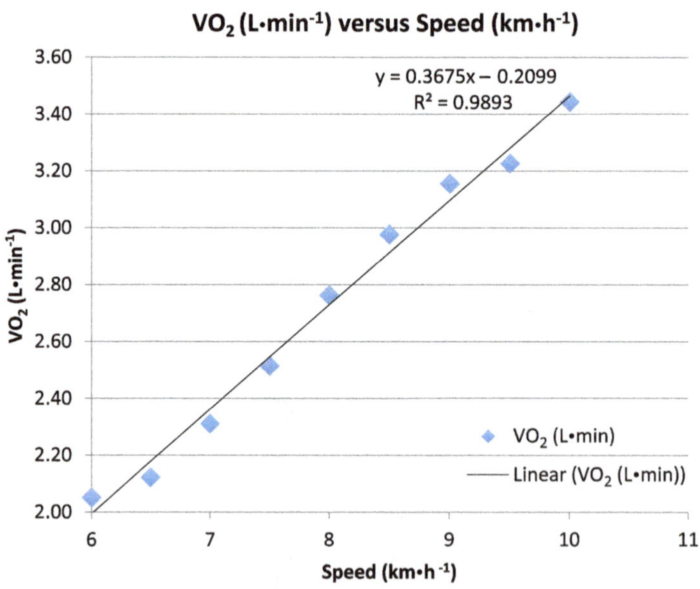

Figure B. Hypothetical graph of the participant's linear regression equation graph.

1.3. Log values for VO_2 (L•min^{-1}) and lactate in Excel table format (logging the values will provide more accurate results).

Sample for VO_2 (L•min^{-1}) in Excel sheet:

= LOG(2.05)

= 0.31

Sample for lactate (mmol•L^{-1}) in Excel sheet:
= LOG(0.88)
= –0.06

Do the same for the rest of the values.

1.4. Find the intersection. Remove outliers and identify the stage which indicates a significant increase in lactate values (Table A.2). Create a second linear series.

Table A.2. Hypothetical participant's data log VO_2 (L•min^{-1}) versus log lactate (mmol•L^{-1}).

VO_2 (L•min^{-1})	0.31	0.33	0.36	0.40	0.44	0.47	0.50	0.51	0.54
Lactate (mmol•L^{-1})	–0.06			0.10	0.18				
						0.35	0.51		0.67
		–0.16	0.08						
								0.58	

Outliers

Intersection/significant increase in lactate values

1.5. Plot the linear regression equation graph. Ensure that R^2 values are as close to 0.99 as possible for both linear series (Figure C).

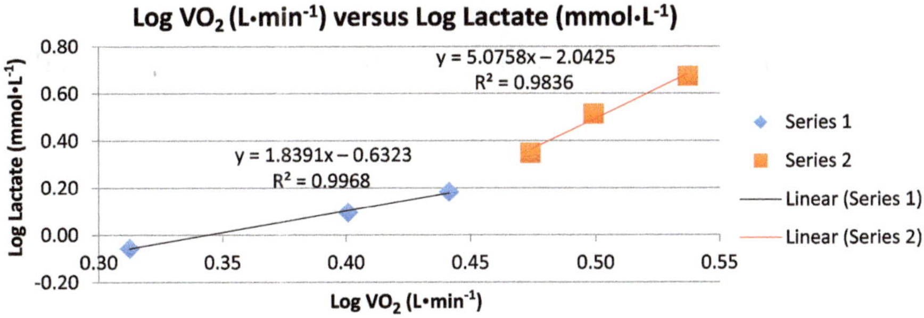

Figure C. Hypothetical linear regression equation graph log blood lactate (mmol•L^{-1}) versus log VO_2 (L•min^{-1}).

1.6. Using simultaneous equation, find the value of x:

$y = 5.0758x - 2.0425$, $y = 1.8391x - 0.6323$

$5.0758x - 2.0425 = 1.8391x - 0.6323$

$5.0758x - 1.8391x = 2.0425 - 0.6323$

$3.2367x = 1.4102$

$x = 1.4102/3.2367$

$x = 0.4356906726$

10^0.4356906726 = 2.7270347532 ("unlog" value)

VO_2 at lactate threshold = 2.73 L•min^{-1}

1.7. Using VO_2 versus speed linear regression equation graph:

Using VO_2 versus speed graph:

Sub y = 2.73 into y = mx + c equation (function)

$y = 0.3675x - 0.2099$

$2.73 = 0.3675x - 0.2099$

$0.3675x = 2.73 + 0.2099$

$0.3675x = 2.9399$

$x = 2.9399/0.3675$

$x = 8.1325$ km•h^{-1}

vLT≈ 8.1 km•h^{-1}

Therefore, the participant's velocity (or speed) at lactate threshold is 8.1 km•h^{-1}.

2. Using VO_2 ($L{\cdot}min^{-1}$) versus VE/VO_2 and VE/VCO_2 ($mL{\cdot}kg^{-1}{\cdot}min^{-1}$)to Detect V_{pt} (Thor & Balasekaran, 2012)

This is a non-log-log graph plot method, but it can be done with a log-log plot method similar to the lactate threshold method (Figure C).

2.1. Average the last 2 minutes of the participant's VO_2 ($L{\cdot}min^{-1}$), VEO_2, and $VECO_2$ ($mL{\cdot}kg^{-1}{\cdot}min^{-1}$) data for each stage (similar to 1.1). Put the values in a table form (Table A.3).

Table A.3. Sample hypothetical data of participant's VO_2 ($L{\cdot}min^{-1}$) versus VE/VO_2 and VE/VCO_2 ($mL{\cdot}kg^{-1}{\cdot}min^{-1}$).

VO_2 ($L{\bullet}min^{-1}$)	2.05	2.12	2.31	2.52	2.76	2.98	3.16	3.23	3.44
VE/VO_2 ($mL{\bullet}kg^{-1}{\bullet}min^{-1}$)	21.93	23.21	24.09	25.41	26.35				
						27.84	29.46	30.19	31.89
VE/VCO_2 ($mL{\bullet}kg^{-1}{\bullet}min^{-1}$)	27.35	27.99	28.47	29.17	29.62				
						30.53	31.58	32.21	33.28

Intersection/significant increase in VO_2 values

Find the intersection. Remove outliers (if any) and identify the stage which indicated the significant increase in VO_2 values. Create a second linear series for VE/VO_2 and VE/VCO_2.

2.2. Plot a linear regression equation graph for the participant's VO_2 ($L{\cdot}min^{-1}$) versus VE/VO_2 and VE/VCO_2 ($mL{\cdot}kg^{-1}{\cdot}min^{-1}$) (Figure D). Ensure that the R^2 value is as close to 0.99 as possible (you may do so by removing outliers).

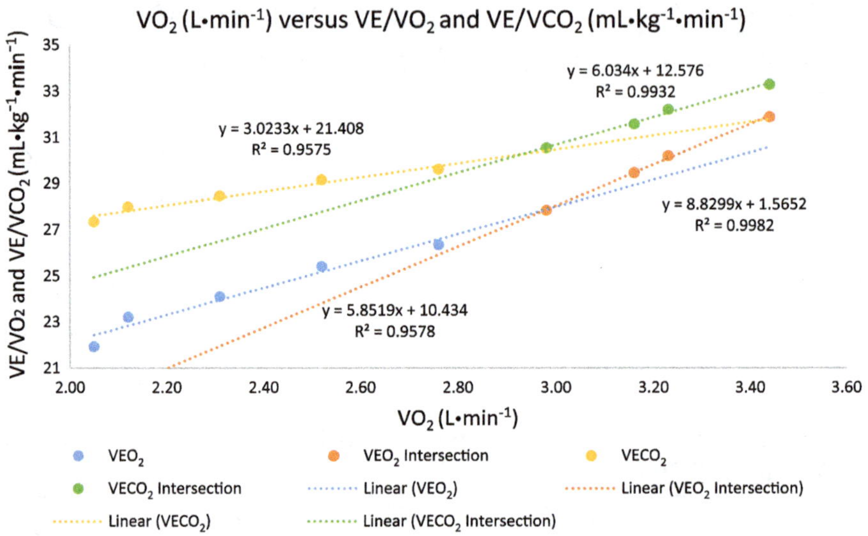

Figure D. Hypothetical linear regression equation graph VO_2 ($L{\bullet}min^{-1}$) versus VE/VO_2 and VE/VCO_2 ($mL{\bullet}kg^{-1}{\bullet}min^{-1}$).

2.3. Using simultaneous equation, solve x for VE/VO_2:

$y = 8.8299x + 1.5652$, $y = 5.8519x + 10.434$

$8.8299x + 1.5652 = 5.8519x + 10.434$

$8.8299x - 5.8519x = 10.434 + 1.5652$

$2.978x = 8.8688$

$x = 8.8688/2.978$

$x = 2.97810611\ L{\bullet}min^{-1}$

$x \approx 2.98\ L{\bullet}min^{-1}$

2.4. Using VO_2 versus speed linear regression equation graph:

Using VO_2 versus speed graph:

Sub $y = 2.98$ into $y = mx + c$ equation (function)

$y = 0.3675x - 0.2099$

$2.98 = 0.3675x - 0.2099$

$0.3675x = 2.98 + 0.2099$

$0.3675x = 3.188$
$x = 3.188/0.3675$
$x = 8.675$
$x \approx 8.7$ km•h^{-1}

2.5. Using simultaneous equation, solve x for VE/VCO_2:
$y = 6.034x + 12.576$, $y = 3.0233x + 21.408$
$6.034x + 12.576 = 3.0233x + 21.408$
$6.034x - 3.0233x = 21.408 - 12.576$
$3.0107x = 8.832$
$x = 8.832/3.0107$
$x = 2.93353705$
$x \approx 2.93$ L•min^{-1}

2.6. Using VO_2 versus speed linear regression equation graph:
Using VO_2 versus speed graph:

Sub $y = 2.93$ into $y = mx + c$ equation (function)

$y = 0.3675x - 0.2099$
$2.93 = 0.3675x - 0.2099$
$0.3675x = 2.93 + 0.2099$
$0.3675x = 3.1399$
$x = 3.1399/0.3675$
$x = 8.544$
$x \approx 8.5$ km•h^{-1}

3. Using Speed versus VE/VO_2 and VE/VCO_2 Method to Detect V_{pt} (Thor & Balasekaran, 2012)

This is a non-log-log graph plot method, but it can be done with a log-log plot method similar to the lactate threshold method (Figure C).

3.1. Average the last 2 minutes of the participant's VE/VO_2 and VE/VCO_2 ($mL \cdot kg^{-1} \cdot min^{-1}$) data over treadmill speed ($km \cdot h^{-1}$) for each stage (similar to 1.1). Put the values in a table form (Table A.4).

Table A.4. Sample hypothetical data of the participant's speed ($km \cdot h^{-1}$) versus VE/VO_2 and VE/VCO_2 ($mL \cdot kg^{-1} \cdot min^{-1}$).

Speed ($km \cdot h^{-1}$)	6	6.5	7	7.5	8	8.5	9	9.5	10
VE/VO_2 ($mL \cdot kg^{-1} \cdot min^{-1}$)	21.93	23.21	24.09	25.41	26.35				
						27.84	29.46	30.19	31.89
VE/VCO_2 ($mL \cdot kg^{-1} \cdot min^{-1}$)	27.35	27.99	28.47	29.17	29.62				
						30.53	31.58	32.21	33.28

Intersection/significant increase in VO_2 values

3.2. Plot a linear regression equation graph for the participant's speed ($km \cdot h^{-1}$) versus VE/VO_2 and VE/VCO_2 ($mL \cdot kg^{-1} \cdot min^{-1}$). Ensure that the R^2 value is as close to 0.99 as possible (you may do so by removing outliers) (Figure E).

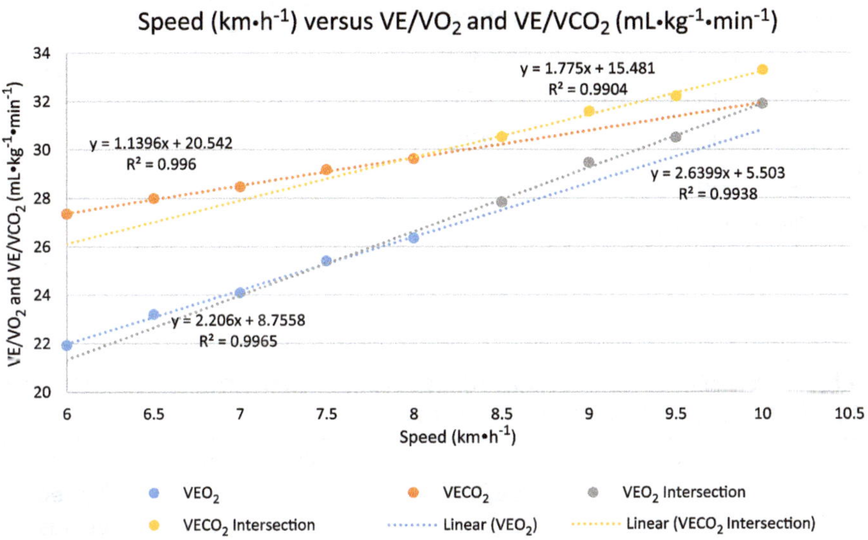

Figure E. Hypothetical linear regression equation graph speed (km•h^{-1}) versus VE/VO_2 and VE/VCO_2 (mL•kg^{-1}•min^{-1}).

3.3. Using simultaneous equation, solve x for VE/VO_2:

$y = 2.6399x + 5.503$, $y = 2.206x + 8.7558$

$2.6399x + 5.503 = 2.206x + 8.7558$

$2.6399x - 2.206x = 8.7558 - 5.503$

$0.4339x = 3.2528$

$x = 3.2528/0.4339$

$x = 7.49665822$

$x \approx 7.5$ km•h^{-1}

3.4. Using simultaneous equation, solve x for VE/VCO_2:

$y = 1.775x + 15.481$, $y = 1.1396x + 20.542$

$1.775x + 15.481 = 1.1396x + 20.542$

$1.775x - 1.1396x = 20.542 - 15.481$

$0.6354x = 5.061$

$x = 5.061/0.6354$

$x = 7.96506138$

$x \approx 8.0$ km•h^{-1}

Table A.5. Overview of speed at lactate threshold or ventilatory threshold (V_{pt}) using various methods.

Method	Log-Log Method	VE/VO_2 (VO_2 versus VE/VO_2)	VE/VCO_2 (VO_2 versus VE/VCO_2)	VE/VO_2 (Speed versus VE/VO_2)	VE/VCO_2 (Speed versus VE/VCO_2)
Speed (km•h^{-1})	8.1	8.7	8.5	7.5	8.0

*Note for above methods:

Method 1 (log-log method) is the most accurate. Using actual results to plot, Table A.5 seems to indicate that methods 2 and 3 have produced slightly higher results to determine speed at lactate threshold and V_{pt}. Gas indices will be accurate if gas calibration for the oxygen uptake machine is accurately calibrated. Lactate will also be accurate if the lactate analyser is calibrated accurately. There may be differences in determining values between the YSI lactate analyser and a portable lactate analyser. By physiological rationale, lactate threshold and V_{pt} have similar results. Therefore, the student should be able to troubleshoot if V_{pt} and lactate threshold detection do not produce similar speed results from the calculations.

4. Heart Rate and Lactate Method versus VO_2 Using Sample Data to Detect HR at V_{pt} (Thor & Balasekaran, 2012)

4.1. Average the last 2 minutes of the participant's VO_2 (mL•kg^{-1}•min^{-1}) data for each stage (similar to 1.1). Put the values in a table form (Table A.6).

Table A.6. Sample hypothetical data of participant's speed (km•hr^{-1}) versus VO_2 (mL•kg^{-1}•min^{-1}).

Speed (km•h^{-1})	6	6.5	7	7.5	8	8.5	9	9.5	10
VO_2 (mL•kg^{-1}•min^{-1})	28.39	29.37	31.97	34.80	38.22	41.18	43.64	44.63	47.64

4.2. Plot a linear regression equation graph for the participant's speed (km•hr^{-1}) versus VO_2 (mL•kg^{-1}•min^{-1}) (Figure F). Ensure that the R^2 value is as close to 0.99 as possible. You may remove outliers to ensure the value of 0.99.

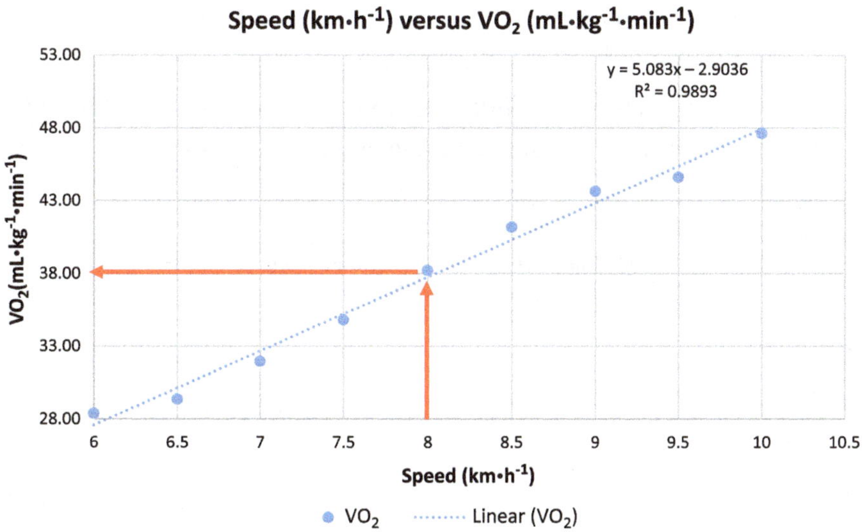

Figure F. Hypothetical linear regression equation graph speed (km•h^{-1}) versus VO_2 (mL•kg^{-1}•min^{-1}).

4.3. Average the last 2 minutes of the participant's VO_2 ($mL{\cdot}kg^{-1}{\cdot}min^{-1}$), heart rate (HR, $beats{\cdot}min^{-1}$), and lactate (LT, $mmol{\cdot}L^{-1}$) data for each stage (similar to 1.1). Put the values in a table form (Table A.7).

Table A.7. Sample hypothetical data of the participant's speed VO_2 ($mL{\cdot}kg^{-1}{\cdot}min^{-1}$) versus heart rate ($beats{\cdot}min^{-1}$) and lactate ($mmol{\cdot}L^{-1}$).

VO_2 ($mL{\cdot}kg^{-1}{\cdot}min^{-1}$)	28.39	29.37	31.97	34.80	38.22	41.18	43.64	44.63	47.64
Heart Rate ($beats{\cdot}min^{-1}$)	132.00	135.50	144.21	151.50	160.25	174.41	180.17	183.75	194.30
Lactate ($mmol{\cdot}L^{-1}$)		0.69		1.25	1.53	2.23	3.27	3.84	4.72
	0.88		1.19						

Outliers

4.4. Plot a graph for the participant's VO_2 ($mL{\cdot}kg^{-1}{\cdot}min^{-1}$) versus heart rate (HR, $beats{\cdot}min^{-1}$) and lactate (LT, $mmol{\cdot}L^{-1}$) (Figure G). Ensure that the R^2 value is as close to 0.99 as possible. You may remove outliers to ensure the value of 0.99. Create a primary and secondary y-axis for HR and LT, respectively. Select the linear regression line for HR and exponential line for LT.

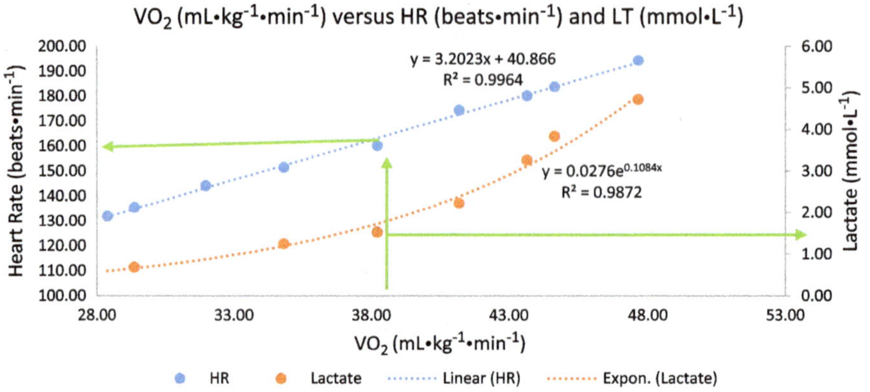

Figure G. Hypothetical graph VO_2 ($mL{\cdot}kg^{-1}{\cdot}min^{-1}$) versus heart rate (HR, $beats{\cdot}min^{-1}$) and lactate (LT, $mmol{\cdot}L^{-1}$).

If we select 8.1 km•hr^{-1} as speed at V_{pt}, we can derive the following:

VO_2 at V_{pt}: 38.22 mL•kg^{-1}•min^{-1} (Figure F)

HR at V_{pt}: 160.25 beats•min^{-1} (Figure G)

≈ 160 beats•min^{-1}

LT at V_{pt}: 1.53 mmol•L^{-1}

Participant's age: 24 years old

Max HR (HR_{max}): 220 – 24 (age) = 196 HR_{max}

160/196 × 100 = 81.6% of HR_{max}

5. Heart Rate and Lactate Threshold Method versus Speed Using Sample Data to Detect Heart Rate at V_{pt} (Thor & Balasekaran, 2012)

5.1. Average the last 2 minutes of the participant's heart rate (HR, beats•min^{-1}) and lactate (LT, mmol•L^{-1}) data for each stage (similar to 1.1). Put the values in a table form (Table A.8).

Table A.8. Sample hypothetical data of the participant's speed (km•hr^{-1}) versus heart rate (HR, beats•min^{-1}) and lactate (LT, mmol•L^{-1}).

Speed (km•h^{-1})	6	6.5	7	7.5	8	8.5	9	9.5	10
Heart Rate (beats•min^{-1})	132.00	135.50	144.21	151.50	160.25	174.41	180.17	183.75	194.30
Lactate (mmol•L^{-1})		0.69		1.25	1.53	2.23	3.27	3.84	4.72
	0.88		1.19						

↑ Outliers

5.2. Plot a graph for the participant's speed (mL•kg^{-1}•min^{-1}) versus heart rate (HR, beats•min^{-1}) and lactate (LT, mmol•L^{-1}). Ensure that the R^2 value is as close to 0.99 as possible. You may remove outliers to ensure the value of 0.99. Create a primary and secondary y-axis for HR and LT, respectively. Select the linear regression line for HR and exponential line for LT.

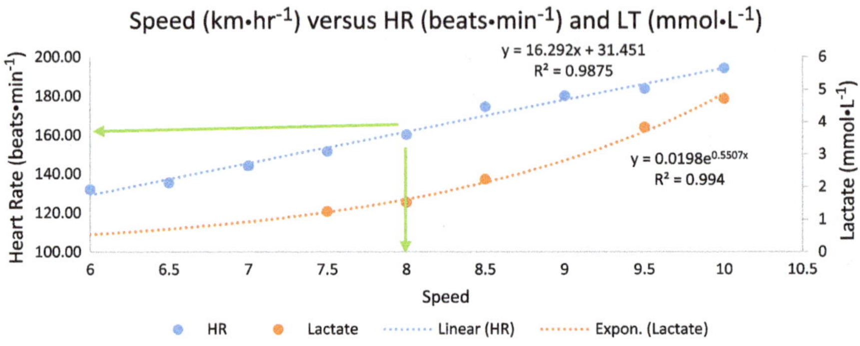

Figure H. Hypothetical graph speed (km•hr^{-1}) versus heart rate (HR, beats•min^{-1}) and lactate (LT, mmol•L^{-1}).

If we select 8.1 km•hr^{-1} as speed at V_{pt}, we can derive the following:

HR at V_{pt}: 160.25 beats•min^{-1} (Figure H)

≈ 160 beats•min^{-1}

LT at V_{pt}: 1.53 mmol•L^{-1}

Participant's age: 24 years old

Max HR (HR_{max}): 220 − 24 (age) = 196 HR_{max}

160/196 × 100 = 81.6% of HR_{max}

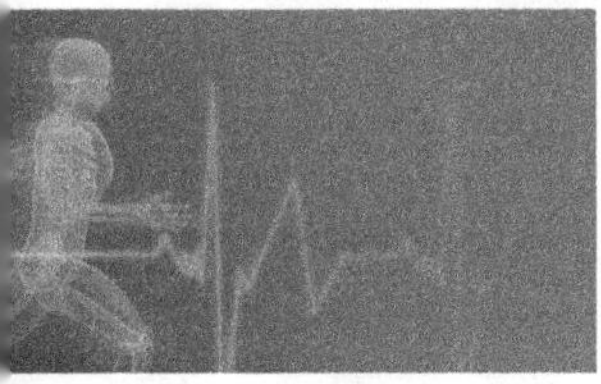

References

1. Ali, M. J., Balasekaran, G., Hoon, K. H., & Gerald, S. (2017). Physiological Differences Between a Non-Continuous and a Continuous Endurance Training Protocol in Recreational Runners and Metabolic Demand Prediction. *Physiological Reports, 1*(1), 30–35.
2. American College of Sports Medicine. (2010). *ACSM's Guidelines for Exercise Testing and Prescription*. Lippincott Williams & Wilkins.
3. American College of Sports Medicine. (2012). *ACSM's Resource Manual for Guidelines for Exercise Testing and Prescription*. Lippincott Williams & Wilkins.
4. American College of Sports Medicine. (2015). *ACSM's Guidelines for Exercise Testing and Prescription*. Wolters & Kluwer.
5. American College of Sports Medicine (Ed.). (2014). *ACSM's Resources for the Health Fitness Specialist*. Lippincott Williams & Wilkins.
6. American College of Sports Medicine (Ed.). (2018). *ACSM's Guidelines for Exercise Testing and Prescription*. Wolters Kluwer.
7. Andreacci, J. L., Robertson, R. J., Dube, J. J., Aaron, D. J., Balasekaran, G., & Arslanian, S.A. (2004). Comparison of Maximal Oxygen Consumption between Black and White Prepubertal and Pubertal Children. *Pediatric Research, 56*(6), 706–713.
8. Arsac, L. M., & Locatelli, E. (2002). Modeling the energetics of 100-m running by using speed curves of world champions. *Journal of Applied Physiology, 92*(5), 1781–1788.
9. Åstrand, P. O. (2003). *Textbook of Work Physiology: Physiological Bases of Exercise* (4th ed.). Human Kinetics.
10. Baechle, T. R., & Earle, R. W. (2000). *Essentials of Strength Training and Conditioning*. Human Kinetics.

11. Balasekaran, G. (1993). To Determine the Exercise Economy of an Experimental Innersole as Compared to that of a Traditional Innersole while Walking, Jogging and Running on a Treadmill. (Masters dissertation, Indiana University of Pennsylvania, United States).
12. Balasekaran, G. (1999). Determination of a Mass Exponent for Maximal Aerobic Power and Peak Anaerobic Power in African-American and Caucasian Children. (PhD dissertation, University of Pittsburgh, United States).
13. Balasekaran, G. (2001). Physiological Factors of Long Distance Running Affected by Biomechanical Efficiency of Human Movement. *International Association of Athletics Federations Bulletin*, 1, 25–35.
14. Balasekaran, G. (2002). The Physiology of Lactate and its Application in Long Distance Running. *International Association of Athletics Federations Bulletin*, 2, 25–30.
15. Balasekaran, G. (2003). Body Composition in Track and Field: Measurement and Application. *International Association of Athletics Federations Bulletin*, 1, 30–35.
16. Balasekaran, G., Boey, P., & Ng, Y. C. (2018). Effects of Self-regulating Exercise Intensity using the OMNI Rate of Perceived Exertion Scale on Youths and Pedagogical Methods for Youths during Physical Education in Singapore. In S. Popoviæ, B. Antala, D. Bjelica, & J. Gardaševiæ (Eds.), *Physical Education in Secondary School: Researches—Best Practices—Situation* (pp. 129–143). Faculty of Sport and Physical Education of University of Montenegro, Montenegrin Sports Academy and Fédération Internationale D´Éducation Physique (FIEP Europe).
17. Balasekaran, G., Boey, P., & Ng, Y. C. (2020). Best Practices in Physical Education and Physical Activity in Nanyang Technological University Singapore. In M. Bobrík, B. Antala, & R. Pělucha (Eds.), Physical Education in Universities: Researches —Best Practices—Situation (pp. 285–293).Slovak Scientific Society for Physical Education and Sport and FIEP.
18. Balasekaran, G., Boey, P., Hui, S. S-C., Govindaswamy, V. V., Ng, Y. C., & Lim, Z. J. (2018). Correlation of Handgrip Strength and Cardiovascular Fitness with Percent Body Fat in Singapore Adolescents. *Gazzetta Medica Italiana—Archivio per le Scienze Mediche, 177*(5), 198–203.

19. Balasekaran, G., Govindaswamy, V., Cheo, N. Y., & Boey, P. (2019). Best Practices in Physical Education in Singapore's Early Childhood Education and Care. In B. Antala, G. Demirhan, A. Carraro, C. Oktar, H. Oz, & A. Kaplanova (Eds.), *Physical Education in Early Childhood Education and Care—Researches, Best Practices, Situation*. (pp. 215–25). Slovak Scientific Society for Physical Education and Sport and FIEP.
20. Balasekaran, G., Gupta, N., & Govindaswamy, V. V. (2010). Assessment Of Body Composition. In M. Chia & J. Chiang (Eds.), *Sport Science & Studies in the East: Issues, Reflections and Emergent Solutions* (pp. 200–220). World Scientific.
21. Balasekaran, G., Gupta, N., Govindaswamy, V. V., Wang, P. K., & Bakri, A. Z. (2014). Physical Education Story: A Journey of Transformations in Singapore. In M. K. Chin & G. Edginton (Eds.), *Physical Education and Health: Global Perspectives and Best Practice* (pp. 409–420). Sagamore Publishing LLC.
22. Balasekaran, G., Ismail, I., & Thor, D. (2015). Effect of 4 weeks of Fun Game-Based and Structured Interval Training Physical Education Lessons on Aerobic Fitness in Adolescents. *Asian Journal of Physical Education and Sport Science*, 5, 1–9.
23. Balasekaran, G., Lim, J. Z., Boey, P., Ng, Y. C., & Govindaswamy, V. V. (2020). AQUATITAN™ lower body compression garment results in lower 200-m run timings. *Gazzetta Medica Italiana-Archivio per le Scienze Mediche, 179*(6), 412–418.
24. Balasekaran, G., Lim, Z. J., Boey, P., Govindaswamy, V. V., Foo, W., & Ng, Y. C. (2019). Acute Foam Rolling on Quadriceps Performance and Short-Term Recovery from Fatigue. *Gazzetta Medica Italiana—Archivio per le Scienze Mediche, 1*(1), 200–229. (RPE)
25. Balasekaran, G., Lim, Z. J., Govindaswamy, V. V., Ee, S. W., & Ng, Y. C. (2019). Effect of AquaTitan Bracelet on Quadriceps Recovery after Fatiguing Muscular Strength and Endurance Exercise. *Gazzetta Medica Italiana—Archivio per le Scienze Mediche, 1*(2), 100–120.
26. Balasekaran, G. & Loh, M. K. (2009). School Physical Education Programmes: Health & Fitness Issues and Challenges. In N. Aplin (Ed.), *An Eye on the Youth Olympic Games 2010: Perspectives on PE and Sport Science in Singapore* (pp. 50–60). McGraw-Hill.

27. Balasekaran, G., Loh, M. K., Govindaswamy, V. V., & Cai, S. J. (2014). OMNI Scale Perceived Exertion Responses in Obese and Normal Weight Male Adolescents During Cycle Exercise. *Journal of Sports Medicine and Physical Fitness, 54*(2), 186–196.
28. Balasekaran, G., Loh, M. K., Govindaswamy, V. V., & Robertson, R. J. (2012). OMNI Scale of Perceived Exertion: Mixed Gender and Race Validation for Singapore Children During Cycle Exercise. *European Journal of Applied Physiology*, 1, 35–38.
29. Balasekaran, G., Loh M. K., Vikneswaran V., Yong T. Z., Govindaswamy V. V., & Ng, Y. C. (2021). Energy System Contribution during 1500m Running in Untrained and Endurance-Trained Asian Male College Students. *The Asian Journal of Kinesiology*, 23(2), 9–18.
30. Balasekaran, G., Mayo, M., & Lim, J. (2019). Fat Distribution and Metabolic Risk Factors of Young Obese Males Following the Cessation of Training: A Follow Up. *Translational Sports Medicine, 2*(2), 82–89.
31. Balasekaran, G., Robertson, R. J., Goss, F. L., Suprasongsin, C., Danadian, K., Govindaswamy V. V., & Arslanian, S. (2005). Short-term Pharmacological Induced Growth Study of Ontogenetic Allometry of Oxygen Uptake in Children. *Annals of Human Biology, 32*(6), 746–759.
32. Balasekaran, G., Robertson, R. J., Loh, M. K., Mayo, M., Lelieveld, A., Grantham, J., & Govindaswamy, V. (2003). Concurrent validity of the OMNI perceived scale using cross modal pictorial descriptors. *Medicine & Science in Sports & Exercise, 35*(5), S58
33. Balasekaran, G., Thor, D., Govindaswamy, V. V., & Ng Y. C. (2014). OMNI Scale of Perceived Exertion: Self-Regulation of Exercise Intensity in Youths and Pedagogical Approaches for Youths in Physical Education in Singapore. *The Asian Journal of Youth Sport, 1*(1), 43–50.
34. Baun, W. B., Baun, M. R., & Raven, P. B. (1981). A nomogram for the estimate of percent body fat from generalized equations. *Research Quarterly for Exercise and Sport, 52*(3), 380–384.
35. Bernard, O., Maddio, F., Ouattara, S., Jimenez, C., Charpenet, A., Melin, B., & Bittel, J. (1998). Influence of the oxygen uptake slow component on the aerobic energy cost of high-intensity submaximal treadmill running in humans. *European Journal of Applied Physiology & Occupational Physiology, 78*(6), 578–585.

36. Billat, L. V. (2001). Interval training for performance: a scientific and empirical practice. *Sports medicine*, *31*(1), 13–31.
37. Billat, V. & Lopes, P. (2006). Indirect methods for estimation of aerobic power. In P. J. Maud & C. Foster (Eds.), *Physiological Assessment of Human Fitness* (pp. 19–38). Human Kinetics.
38. BMI, O. C. (1998). Clinical guidelines on the identification, evaluation, and treatment of overweight and obesity in adults.
39. Boyd, E. (1935). *The growth of the surface area of the human body*. University of Minnesota Press.
40. Brown, S. A., Chambless, L. E., Sharrett, A. R., Gotto Jr, A. M., & Patsch, W. (1992). Postprandial lipemia: reliability in an epidemiologic field study. *American journal of epidemiology*, *136*(5), 538–545.
41. Brown, S. P., Miller, W. C., & Eason, J. M. (2006). *Exercise physiology: Basis of human movement in health and disease*. Lippincott Williams & Wilkins.
42. Brozek, J. (1959). Body measurements, including skinfold thickness, as indicators of body composition. *Techniques for measuring body composition*, 3–35.
43. Bruce, R., Kusumi, F., & Hosmer, D. (1973). Maximal oxygen intake and nomographic assessment of functional aerobic impairment in cardiovascular disease. *American Heart Journal*, *85*(4), 546–562.
44. Bundle, M. W., Hoyt, R. W., & Weyand, P. G. (2003). High-speed running performance: a new approach to assessment and prediction. *Journal of Applied Physiology*, *95*(5), 1955–1962.
45. Cai, S. J. & Balasekaran, G. (2004). Validation of OMNI Rate of Perceived Exertion Scale in Obese Male Individuals. (Honours dissertation, National Institute of Education, Nanyang Technological University, Singapore).
46. Chia, J., & Balasekaran, G. (2018). Rate of Perceived Exertion (RPE)—A Safe Self-Regulation of Exercise Intensity for Children in Singapore. (Masters dissertation, National Institute of Education, Nanyang Technological University, Singapore).
47. Ching, B. & Horsfall, P. A. (1977). Lung volumes in normal Cantonese subjects: preliminary studies. *Thorax*, *32*(3), 352–355.
48. Cohen, J. C., Noakes, T. D., & Benade, A. S. (1989). Postprandial lipemia and chylomicron clearance in athletes and in sedentary men. *The American Journal of Clinical Nutrition*, *49*(3), 443–447.

49. Danadian, K., Balasekaran, G., Lewy, V., Meza, M. P., Roberston, R. J., & Arslanian, S. A. (1999). Insulin Sensitivity in African-American Children With and Without Family History of Type 2 Diabetes. *Diabetes Care, 22*(8), 1325–1329.
50. Demura, S., Yamaji, S., & Kitabayashi, T. (2006). Residual volume on land and when immersed in water: Effect on percent body fat. *Journal of Sports Sciences, 24*(8), 825–833.
51. Di Prampero, P. E. (1986). The energy cost of human locomotion on land and in water. *International Journal of Sports Medicine*, 7(2), 55–72.
52. Di Prampero, P. E. (2003). Factors limiting maximal performance in humans. *European Journal of Applied Physiology, 90*(3–4), 420–429.
53. Di Prampero, P. E., Capelli, C., Pagliaro, P., Antonutto, G., Girardis, M., Zamparo, P., & Soule, R. G. (1993). Energetics of best performances in middle-distance running. *Journal of Applied Physiology, 74*(5), 2318–2324.
54. Di Prampero, P. E., Fusi, S., Sepulcri, L., Morin, J. B., Belli, A., & Antonutto, G. (2005). Sprint running: A new energetic approach. *Journal of Experimental Biology, 208*(14), 2809–2816.
55. Dill, D. B. & Costill, D. L. (1974). Calculation of percentage changes in volumes of blood, plasma, and red cells in dehydration. *Journal of Applied Physiology, 37*(2), 247–248.
56. Dubois, C., Armand, M., Azais-Braesco, V., Portugal, H., Pauli, A. M., Bernard, P. M., Latgé, C., Lafont, H., & Lairon, D. (1994). Effects of moderate amounts of emulsified dietary fat on postprandial lipemia and lipoproteins in normolipidemic adults. *The American Journal of Clinical Nutrition, 60*(3), 374–382.
57. Dubois, C., Beaumier, G., Juhel, C., Armand, M., Portugal, H., Pauli, A. M., Borel, P., Latgé, C., & Lairon, D. (1998). Effects of graded amounts (0–50 g) of dietary fat on postprandial lipemia and lipoproteins in normolipidemic adults. *The American Journal of Clinical Nutrition, 67*(1), 31–33.
58. Dubois, D. & Dubois, E. F. (1916). A formula to estimate the approximate surface area if height and weight is known. *Arch Int Med*, 17, 863–871.
59. Dubois, E. F. (1936). *Basal metabolism in health and disease*. Lea & Febiger.
60. Duffield, R. J., & Dawson, B. (2003). Energy system contribution in track running. *New studies in Athletics, 18*(4), 47–56.

61. Fielding, B. A., Callow, J., Owen, R. M., Samra, J. S., Matthews, D. R., & Frayn, K. N. (1996). Postprandial lipemia: The origin of an early peak studied by specific dietary fatty acid intake during sequential meals. *The American Journal of Clinical Nutrition*, *63*(1), 36–41.
62. Fleisch, A. (1951). Le metabolisme basal standard et sa determination au moyen du'Metabocalculator'. *Helv Med Acta, 18*(1), 23–44.
63. Food and Agriculture Organization, World Health Organization, and United Nations University. (1985). Energy and protein requirements. World Health Organization, Technical Report Ser. 724, p. 71–79.
64. Fukagawa, N. K., Bandini, L. G., & Young, J. B. (1990). Effect of age on body composition and resting metabolic rate. *American Journal of Physiology-Endocrinology and Metabolism, 259*(2), E233–E238.
65. Gehan, E. A., George, S. L. (1970). Estimation of human body surface area from height and weight. *Cancer Chemother Rep,* 54, 225–235.
66. Gerrior, S., Juan, W., & Peter, B. (2006). An easy approach to calculating estimated energy requirements. *Preventing Chronic Disease*, *3*(4).
67. Golding, L. A., Myers, C. R., & Sinning, W. E. (1989). *Y's way to physical fitness: The complete guide to fitness testing and instruction*. YMCA of the USA.
68. Goldman, H. I. & Becklake, M. R. (1959). Respiratory function tests, normal values at median altitudes and the prediction of normal results. *American Review of Tuberculosis and Pulmonary Diseases*, *79*(4), 457–467.
69. Govindasamy B., Govindaswamy, V. V., Loh, R. M. K., Ng, Y. C., Thor, D., Boey, P., & Lim, J. (2019). *Curricular Guide to Health-Fitness Applications in Physical Education Using the OMNI Perceived Exertion Scale*. Pearson Education South Asia Pte Ltd.
70. Graham, K. S. & McLellan, T. M. (1989). Variability of time to exhaustion and oxygen deficit in supramaximal exercise. *Australian Journal of Science and Medicine in Sport, 21*(4): 11–14.
71. Grantham, J. R. & Balasekaran, G. (2004). Effects of Aerobic and Strength Training on Basal Metabolic Rate and Postprandial Lipaemia. (PhD dissertation, National Institute of Education, Nanyang Technological University, Singapore).
72. Grantham, J. R., Mayo, M. J., O'Brien, B. J., & Balasekaran, G. (2004). The effect of acute exercise upon postprandial lipemia following correction for plasma volume shifts. *Medicine & Science in Sports & Exercise*, *36*(5), S217.

73. Gupta, N. & Balasekaran, G. (2013). Running Energy Reserve Index: Mapping, Assessment and Prediction. (PhD dissertation, National Institute of Education, Nanyang Technological University, Singapore).
74. Gupta, N., Balasekaran, G., & Govindaswamy, V. V. (2011). Comparison of Treadmill Protocols to Measure Maximal Oxygen Uptake (VO_{2max}). *Medicine & Science in Sports & Exercise, 43*(5), 730.
75. Gupta, N., Balasekaran, G., Govindaswamy, V. V., Chia, Y. H., & Lim, M. S. (2011). Comparison of body composition with bioelectric impedance (BIA) and dual energy X-ray absorptiometry (DEXA) among Singapore Chinese. *Journal of Science and Medicine and Sport, 14*(1), 33–35.
76. Hardman, A. E. & Herd, S. L. (1998). Exercise and postprandial lipid metabolism. *Proceedings of the Nutrition Society, 57*(1), 63–72.
77. Hartung, G. H., Lawrence, S. J., Reeves, R. S., & Foreyt, J. P. (1993). Effect of alcohol and exercise on postprandial lipemia and triglyceride clearance in men. *Atherosclerosis, 100*(1), 33–40.
78. Haugen, H. A., Chan, L. N., & Li, F. 2007). Indirect Calorimetry: A Practical Guide for Clinicians. *Nutrition in Clinical Practice, 22*(4), 337–388.
79. Hawley, J. A., Burke, L. M., Angus, D. J., Fallon, K. E., Martin, D. T., & Febbraio, M. A. (2000). Effect of altering substrate availability on metabolism and performance during intense exercise. *British Journal of Nutrition, 84*(6), 829–838.
80. Haycock, G. B., Schwartz, G. J., Wisotsky, D. H. (1978). Geometric method for measuring body surface area: A height weight formula validated in infants, children and adults. *The Journal of Pediatrics*, 93(1), 62–66.
81. Hill, A. V. (1925). The physiological basis of athletic records. *The Scientific Monthly, 21*(4), 409–428.
82. Hill, A. V. (1950). The dimensions of animals and their muscular dynamics. *Science Progress (1933-), 38*(150), 209–230.
83. Hoffman, J. R. & Kang, J. (2002). Evaluation of a new anaerobic power testing system. *Journal of Strength and Conditioning Research, 16*(1), 142–148.
84. Howley, E. T. & Franks, B. D. (1986). *Health/Fitness Instructor's Handbook*. Human Kinetics Publishers, Inc.
85. Jackson, A. S., Pollock, M. L., & Gettman, L. R. (1978). Intertester reliability of selected skinfold and circumference measurements and percent fat

estimates. *Research Quarterly. American Alliance for Health, Physical Education and Recreation*, *49*(4), 546–551.

86. Jackson, A. S., Pollock, M. L., & Ward, A. N. N. (1980). Generalized equations for predicting body density of women. *Medicine and Science in Sports and Exercise*, *12*(3), 175–181.
87. Janssen, P. G. J. M. (1994). *Training Lactate Pulse Rate* (4th ed.). Polar Electro.
88. Jensen, J., Jacobsen, S. T., Hetland, S., & Tveit, P. (1997). Effect of combined endurance, strength and sprint training on maximal oxygen uptake, isometric strength and sprint performance in female elite handball players during a season. *International Journal of Sports Medicine*, *18*(5), 354–358.
89. Joyner, M. J. (1991). Modeling: optimal marathon performance on the basis of physiological factors. *Journal of Applied Physiology*, *70*(2), 683–687.
90. Lacour, J. R., Bouvat, E., & Barthelemy, J. C. (1990). Post-competition blood lactate concentrations as indicators of anaerobic energy expenditure during 400-m and 800-m races. *European Journal of Applied Physiology and Occupational Physiology*, *61*(3–4), 172–176.
91. Lacour, J. R., Padilla-Magunacelaya, S., Chatard, J. C., Arsac, L., & Barthelemy, J. C. (1991). Assessment of running velocity at maximal oxygen uptake. *European Journal of Applied Physiology & Occupational Physiology*, *62*(2), 77–82.
92. Lee, M. K. & Balasekaran, G. (2010). The Measurement of Three-Dimensional Body Segment Parameters Using Dual Energy X-Ray Absorptiometry and Skin Geometry an Application in Gait Analysis. (PhD dissertation, National Institute of Education, Nanyang Technological University, Singapore).
93. Loh, M. K. & Balasekaran, G. (2004). Validation of Gender and Racially Specific OMNI Perceived Exertion Rating Scales for Children in Singapore. (Masters dissertation, National Institute of Education, Nanyang Technological University, Singapore).
94. Mackenzie, B. (1997). Energy Pathways. BrianMac Sports Coach.
95. McArdle, W. D., Katch, F. I., & Katch, V. L. (2010). *Exercise Physiology: Nutrition, Energy, and human performance*. Lippincott Williams & Wilkins.
96. Medbø, J. I., Mohn, A. C., Tabata, I., Bahr, R., Vaage, O., and Sejersted, O. M. (1988). Anaerobic capacity determined by maximal accumulated O_2 deficit. *Journal Applied Physiology*, *64*(1): 50–60.

97. Melzer, K. (2011). Carbohydrate and fat utilization during rest and physical activity. *e-SPEN, the European e-Journal of Clinical Nutrition and Metabolism*, *6*(2), e45–e52.
98. Mero, N., Van Tol, A., Scheek, L. M., Van Gent, T., Labeur, C., Rosseneu, M., & Taskinen, M. R. (1998). Decreased postprandial high density lipoprotein cholesterol and apolipoproteins AI and E in normolipidemic smoking men: Relations with lipid transfer proteins and LCAT activities. *Journal of Lipid Research*, *39*(7), 1493–1502.
99. Metropolitan Life Insurance Company. (1983). 1983 Metropolitan Height and Weight Tables. *Statistical Bulletin of the Metropolitan Life Insurance Company 64* (Jan–June): 3.
100. Ministry of Health. My Health Plate. Retrieved from https://www.healthhub.sg/programmes/55/my-healthy-plate
101. Mougios, V., Kazaki, M., Christoulas, K., Ziogas, G., & Petridou, A. (2006). Does the intensity of an exercise programme modulate body composition changes? *International Journal of Sports Medicine*, *27*(3), 178–181.
102. Noakes, T. (2003). *Lore of Running*. Human Kinetics.
103. Paksaichola, A., Chaturapanicha, G., Komoltrib, C., Srikueaa, R., & Pholpramoola, C. (2014). Residual lung volume of female Thai adults. *SCIENCEASIA*, *40*(3), 219–223.
104. Péronnet, F. & Thibault, G. (1989). Mathematical analysis of running performance and world running records. *Journal of Applied Physiology*, *67*(1), 453–465.
105. Pfeiffer, K. A., Pivarnik, J. M., Womack, C. J., Reeves, M. J., & Malina, R. M. (2002). Reliability and validity of the Borg and OMNI rating of perceived exertion scales in adolescent girls. *Medicine and science in sports and exercise*, *34*(12), 2057–2061.
106. Plowman, S. A. & Smith D. L. (1997). *Exercise Physiology for Health, Fitness, and Performance*. 2nd edition. Allyn and Bacon.
107. Poehlman, E. T. (1989). A review: Exercise and its influence on resting energy metabolism in man. *Medicine and Science in Sports and Exercise*, *21*(5), 515–525.
108. Powers, S. K., Howley, E. T. (2009). *Exercise Physiology: Theory and Application to Fitness and Performance*. 7th Edition. McGraw-Hill.

109. Ramsbottom, R., Nevill, A. M., Nevill, M. E., Newport, S., & Williams, C. (1994). Accumulated oxygen deficit and short-distance running performance. *Journal of Sports Sciences, 12*(5), 447–453.

110. Reis, V. M., & Miguel, P. P. (2007). Changes in the accumulated oxygen deficit and energy cost of running 400 metres. *New Studies in Athletics, 22*(2), 49.

111. Renoux, J. C., Petit, B., Billat, V., & Koralsztein, J. P. (1999). Oxygen deficit is related to the exercise time to exhaustion at maximal aerobic speed in middle distance runners. *Archives of Physiology and Biochemistry, 107*(4), 280–285.

112. Riebe, D., Ehrman, J. K., Liguori, G., Magal, M., & American College of Sports Medicine (Eds.). (2018). *ACSM's Guidelines for Exercise Testing and Prescription*. Wolters Kluwer.

113. Riechman, S. E., Zoeller, R. F., Balasekaran, G., Goss, F. L. & Robertson, R. J. (2002). Prediction of 2000 M Rowing Performance in Females Using Indices of Anaerobic and Aerobic Power. *Journal of Sports Science, 20*(9), 681–687.

114. Robergs, R. A. (1991). Nutrition and exercise determinants of postexercise glycogen synthesis. *International Journal of Sport Nutrition and Exercise Metabolism, 1*(4), 307–337.

115. Robertson, R. J. (2004). *Perceived exertion for practitioners: Rating effort with the OMNI picture system*. Human Kinetics.

116. Robertson, R. J., Goss, F. L., Auble, T. E., Cassinelli, D. A., Spina, R. J., Glickman, E. L., Galbreath, R. W., Silberman, R. M., & Metz, K. F. (1990). Cross-modal exercise prescription at absolute and relative oxygen uptake using perceived exertion. *Medicine and science in sports and exercise, 22*(5), 653–659.

117. Robertson, R. J., Goss, F. L., Boer, N., Gallagher, J. D., Thompkins, T., Bufalino, K., Balasekaran, G., Meckes, C., Pintar, J., & Williams, A. (2001). OMNI scale perceived exertion at ventilatory breakpoint in children: response normalized. *Medicine and science in sports and exercise, 33*(11), 1946–1952.

118. Robertson, R. J., Goss, F. L., Boer, N. F., Peoples, J. A., Foreman, A. J., Dabayebeh, I. M., Millich, N. B., Balasekaran, G., Riechman, S. E., Gallagher, J. D., & Thomkins, T. (2000). Validation of the Omni Perceived Exertion Scale for Children Using A Mixed Gender/Race Cohort. *Medicine and Science in Sport and Exercise, 32*(3), 452–458.

119. Robertson, R. J., Goss, F. L., Boer, N. F., Peoples, J. A., Foreman, A. J., Dabayebeh, I. M., Millich, N. B., Balasekaran, G., Riechman, S. E., Gallagher, J. D. & Thompkins, T. (2000). Children's OMNI scale of perceived exertion: Mixed gender and race validation. *Medicine & Science in Sports & Exercise, 32*(2), 452.
120. Robertson, R. J., Goss, F. L., Boer, N., Gallagher, J. D., Thompkins, T., Bufalino, K., Balasekaran, G., MeCkes, C., Pintar, J., & Williams, A. (2001). Omni Scale Perceived Exertion at Ventilatory Breakpoint in Children: Response Normalized. *Medicine and Science in Sport and Exercise, 33*(11), 1946–1952.
121. Romon, M., Le Fur, C., Lebel, P., Edme, J. L., Fruchart, J. C., & Dallongeville, J. (1997). Circadian variation of postprandial lipemia. *The American Journal of Clinical Nutrition, 65*(4), 934–940.
122. Sady, S. P., Thompson, P. D., Cullinane, E. M., Kantor, M. A., Domagala, E., & Herbert, P. N. (1986). Prolonged exercise augments plasma triglyceride clearance. *Jama, 256*(18), 2552–2555.
123. Savaglio, S. & Carbone, V. (2000). Human performance: Scaling in athletic world records. *Nature, 404*(6775), 244.
124. Siri, W. E. (1961). Body composition from fluid spaces and density: Analysis of methods. *Nutrition, 9*(5), 480–491.
125. Slaughter, M. H., Lohman, T. G., Boileau, R., Horswill, C. A., Stillman, R. J., Van Loan, M. D., & Bemben, D. A. (1988). Skinfold equations for estimation of body fatness in children and youth. *Human Biology, 60*(5), 709–723.
126. Sparling, P. B., O'Donnell, E. M., & Snow, T. K. (1998). The gender difference in distance running performance has plateaued: An analysis of world rankings from 1980 to 1996. *Medicine and Science in Sports and Exercise, 30*(12), 1725–1729.
127. Spencer, M. R., Gastin, P. B., & Payne, W. (1996). Energy system contribution during 400 to 1500 metres running. *New Studies in Athletics*, 11(4), 59–66.
128. Spencer, M. R., & Gastin, P. B. (2001). Energy system contribution during 200- to 1500-m running in highly trained athletes. *Medicine Science in Sports and Exercise, 33*(1), 157–162.
129. Spriet, L. L. (1995). Anaerobic metabolism during high-intensity exercise. *Exercise Metabolism*, 1–39.

130. Svedenhag, J. & Sjodin, B. (1984). Maximal and submaximal oxygen uptakes and blood lactate levels in elite male middle- and long-distance runners. *International Journal of Sports Medicine, 5*(5), 255–261.
131. Table, M. (2005). *Dietary Reference Intakes for Energy, Carbohydrate, Fiber, Fat, Fatty Acids, Cholesterol, Protein, And Amino Acids* (Vol. 5, pp. 589–768). National Academy Press.
132. Thor, D. & Balasekaran, G. (2012). Comparison of cross-modal omni scale of perceived exertion at ventilatory breakpoint and self-regulated exercises in male adolescents in Singapore (Masters dissertation, National Institute of Education, Nanyang Technological University).
133. Tsetsonis, N. V., Hardman, A. E., & Mastana, S. S. (1997). Acute effects of exercise on postprandial lipemia: A comparative study in trained and untrained middle-aged women. *The American Journal of Clinical Nutrition, 65*(2), 525–533.
134. Utter, A. C., Robertson, R. J., Green, J. M., Suminski, R. R., McAnulty, S. R., & Nieman, D. C. (2004). Validation of the Adult OMNI Scale of perceived exertion for walking/running exercise. *Medicine and Science in Sports and Exercise, 36*(10), 1776–1780.
135. Van Beaumont, W. (1972). Evaluation of hemoconcentration from hematocrit measurements. *Journal of Applied Physiology, 32*(5), 712–713.
136. Verheijen, R. (1998). *The Complete Handbook of Conditioning for Soccer*. Reedswain Inc.
137. Vikneswaran, V. & Balasekaran, G. (2003). Energy System Contribution During 1500-M Running In Untrained Trained College Aged Males. (Honours dissertation, National Institute of Education, Nanyang Technological University, Singapore).
138. Volkov, N. I. & Lapin, V. I. (1979). Analysis of the velocity curve in sprint running. *Medicine and Science in Sports, 11*(4), 332.
139. Van Ingen Schenau, G. J., Jacobs, R., & de Koning, J. J. (1991). Can cycle power predict sprint running performance? *European Journal of Applied Physiology and Occupational Physiology, 63*(3–4), 255–260.
140. Ward-Smith, A. J. (1999). The bioenergetics of optimal performances in middle-distance and long-distance track running. *Journal of Biomechanics, 32*(5), 461–465.

141. Weyand, P. G., & Bundle, M. W. (2005). Energetics of high-speed running: integrating classical theory and contemporary observations. *American Journal of Physiology-Regulatory, Integrative and Comparative Physiology, 288*(4), R956–R965.
142. Weyand, P. G., Lee, C. S., Martinez-Ruiz, R., Bundle, M. W., Bellizzi, M. J., & Wright, S. (1999). High-speed running performance is largely unaffected by hypoxic reductions in aerobic power. *Journal of Applied Physiology*, *86*(6), 2059–2064.
143. Wideman Jr, R. F., Kirby, Y. K., Tackett, C. D., Marson, N. E., & McNew, R. W. (1996). Cardio-pulmonary function during acute unilateral occlusion of the pulmonary artery in broilers fed diets containing normal or high levels of arginine-HCl. *Poultry Science*, *75*(12), 1587–1602.
144. Wilson, W. C., Grande, C. M., & Hoyt, D. B. (2007). *Trauma: Critical Care*. CRC Press.
145. Withers, R. T., Sherman, W. M., Clark, D. G., Esselbach, P. C., Nolan, S. R., Mackay, M. H., & Brinkman, M. (1991). Muscle metabolism during 30, 60 and 90s of maximal cycling on an air-braked ergometer. *European Journal of Applied Physiology, 63*, 354–362.
146. World Health Organization. (2004). Expert consultation: Appropriate body-mass index for Asian populations and its implications for policy and intervention strategies. *Lancet*, 363, 157–163.
147. World Health Organization. (2008). WHO STEP wise approach to surveillance (STEPS).
148. World Health Organization. (2011). Waist circumference and waist-hip ratio: Report of a WHO expert consultation, Geneva, 8-11 December 2008.
149. Yap, W. S., Chan, C. C., & Chan, S. P. (2001). Ethnic differences in anthropometry among adult Singaporean Chinese, Malays and Indians, and their effects on lung volumes. *Respiratory Medicine*, *95*(4), 297–304.
150. Yong, T. W. & Balasekaran, G. (2003). Energy System Contribution During 1500-M Running In Untrained College Aged Males And Females. (Honours dissertation, National Institute of Education, Nanyang Technological University, Singapore)
151. Zuntz, N. (1901). Ueber die Bedeutung der verschiedenen Nährstoffe als Erzeuger der Muskelkraft. *Archiv für die gesamte Physiologie des Menschen und der Tiere*, *83*(10–12), 557–571.

Index

CPSIA information can be obtained
at www.ICGtesting.com
Printed in the USA
BVHW050318110921
616351BV00001B/4